A PRIMER on SIMULATION and GAMING

A PRIMER on SIMULATION and GAMING

RICHARD F. BARTON
Director of Planning and Analyses
Professor of Management and Computer Science
Texas Tech University

PRENTICE-HALL, INC., ENGLEWOOD CLIFFS, NEW JERSEY

C—13-700831-7
P—13-700823-6

Library of Congress Catalog Card Number: 79-110489

Current printing (last number)
10 9 8 7 6 5 4 3 2 1

Prentice-Hall International, Inc., *London*
Prentice-Hall of Australia Pty. Ltd., *Sydney*
Prentice-Hall of Canada, Ltd., *Toronto*
Prentice-Hall of India (Private) Ltd., *New Delhi*
Prentice-Hall of Japan, Inc., *Tokyo*

To Nancy

PREFACE

This book is an introduction to simulation and gaming for the administrative professions, for the behavioral sciences, and for education.

Simulation as a method is rapidly pervading decision making, research, and training of all kinds. It has different meanings and uses in different fields. In this book, these diverse views and uses are brought together into a single presentation. This wide purview and the book's purpose as an introduction force the elimination of much specific detail from each field. However, readers should gain sufficient understanding and appreciation of simulation that, with the specifics from their own fields, they will be able to organize and execute actual simulations for themselves.

Computers are not a necessity for running simulations. A reader versed in the theory and practice of his own field can, with this book, organize his own noncomputer man-model simulations or gaming exercises.

For computer applications, this book does not teach

any particular computer language. Programs are shown in the form of flow diagrams or examples. If a reader knows (or learns) some computer language, this book will prepare him to undertake man-computer and all-computer simulations for himself.

For those who do not wish to run simulations, this book offers an understanding of what is going on when others "simulate" or "game."

Computers have become a new frontier for administrative, behavioral, and educational fields. For practitioners and scholars in these areas, the simultaneous sharing of a large computer by many users and the development of user-oriented languages offer a new way of life. The adaptation of these fields to computer-related technological development has only just begun. This book outlines time-sharing simulation and gaming and orients the reader to present and future potentials.

Many persons and a wide literature have contributed to the writing of this book. First drafts of Chapters I-VI were written while the author was a Research Associate, Center for Regional Studies, University of Kansas. Final drafts were prepared while the author was Director of Planning and Analyses, Texas Tech University, where exemplary manuscript chores were performed by Madeline Butts.

RICHARD F. BARTON
Lubbock, Texas
January, 1970

CONTENTS

A PRIMER on SIMULATION and GAMING

1

INTRODUCTION

To simulate means *to give the appearance of* something else. A simulation, although it is a "thing" in itself, is meaningful to its creators and to its users only in terms of other things. To simulate also means *to give the effect of something else*—so that the meaning and usefulness of a simulation is not merely in a visual or sensory likeness but in a similarity of ideas or a conceptual likeness as well.

There are many kinds of simulation that serve a variety of purposes, but in every instance, the meaning of the simulation is always related to something else. For example, a tree frog changes the color of its skin to give the appearance of its resting place. This frog camouflages itself by simulating its surroundings. Here the simulation is a visual one; it is meaningful and useful only in terms of other things, in this case the green leaf or brown bark on which the frog rests.

An astronaut in training practices space travel in

a simulated capsule that gives the effect of being in space. Space simulation exercises have meaning and usefulness only in terms of actual space flight, which is clearly something else again from riding in a capsule in a laboratory here on earth. Once, while an astronaut was aloft, an earth-bound simulation was performed to solve a capsule emergency occurring in space. The results of the simulation study on earth provided instructions that were radioed to the spacecraft to overcome the emergency.

War games, conducted at great expense by armed forces, are simulations that have meaning only in terms of something else—actual war, and in fact, due to technological advances, a kind of actual war that has never existed. War games are simulations because the sides are not real enemies and because casualties and destruction are only symbolically represented.

Computer programs that trace the inputs and outputs of industries in an economy are simulation studies that give the effect of flows of goods and services and of changes in these flows. Irving Fisher, an economist, once tried to study an economic marketplace by means of a physical model made of tubes, cisterns, stoppers, and levers that governed equilibrium water levels.[1] The flow of water through this device attempted to give the appearance of the operation of a market. The pulsing of electronic signals within the central processing unit of a computer and Fisher's flow of water from cistern to cistern are both simulations, but both are meaningful and useful only because they are about something else.

A frequent example of simulation is the "flight" of a model aircraft in a wind tunnel. Like war games, this example illustrates an important point about the usefulness of simulations; that is, *what the simulation is about may not yet exist*. Why fly a model aircraft in a wind tunnel if a real one similar to the model already exists? One could merely fly the real one, assuming it is safe enough. Aircraft models are flown in wind tunnels to study the effects on real aircraft that many times do not yet exist. Indeed, because of the results of the simulation study, they may never exist. The effect (or operating appearance, so to speak) of a proposed aircraft design may be rejected as undesirable entirely on the basis of the simulation study, so that *what is simulated may never come into being*.

[1] Irving Fisher, *Mathematical Investigations in the Theory of Value and Prices* (New Haven: Yale University Press, 1925), pp. 11–119.

Except for the tree frog, the examples of simulation given above illustrate a mixture of purposes. On the one hand, a simulation study is a way of trying out plans and designs before they are put into actual operation or production. This is the very practical application of simulation as it is used by industry, government, and military branches to help make new decisions or to train decision makers for known systems.

On the other hand, a simulation study may provide new knowledge of the "something else" the study is about. This is the scientific application of simulation.

These applications of simulation are particularly important when the "real thing" cannot be studied directly because it does not yet exist, or is not available, or to work with it directly is too dangerous or too expensive.

Simulation has made significant contributions to both theory and practice. Discoveries about aircraft in flight have been made in wind tunnels. New insights into the behavior of men under stress have come from war games. Facts about man's ability to perceive and act on various kinds of data displays are obtained in simulations of space flight. Warehouses have been located on the basis of simulated shipments. Bus terminals have been designed as a consequence of simulations of commuters waiting in line for transportation. Flaws in civil emergency procedures have been discovered by simulation. Complex interactions never thought of before are discovered when computers execute simulation programs of massive interdependent systems such as a business firm, an economy, a flight to the moon and back, a whole war, or a massive communications network.

A Simple Example

Let us illustrate simulation with a simple game. This game displays certain basic elements of human interaction and of uncertainty.

Some readers may look on this example game as a form of gambling. Even so, such an example is an apt one because the "science of gambling" is extremely relevant in one widely used simulation method called, appropriately, Monte Carlo techniques. Probability theory, so important in the design of experiments including simulation experiments, arose from inquiries into games of chance. Moreover, abrogation to chance is an accepted decision-making technique as evidenced by the assignment of goals to football teams by coin tossing.

Our simple example game is called matching pennies. There are two players: Player R and Player C (later these player names will mean "row" and "column"). In this game the players make simultaneous moves by each placing a penny on a flat surface such as a table or the back of a hand. The payoff rules are given below and are labeled in terms of the outcomes to Player R:

R wins: If the exposed sides of the two pennies match, i.e. if both are heads or both are tails, Player R wins $1,000 from Player C.

R loses: If the exposed sides do not match, i.e., one is a head and the other is a tail, Player C wins $1,000 from Player R.

Assume that each player starts with $100,000 and that the game will end either when one player has lost all his money (and, of course, the other player has $200,000) or after 101 trials. A single trial in this game is defined as the simultaneous exposure of a penny by each player.

In this example, the rules of the game govern one kind of human conflict. It is a rather simple conflict situation, but it does portray some real human behavior most of us have seen at one time or another (except perhaps for the starting wealth of each player and the size of the payoff).

We are interested both in individual human behavior and in the whole system of simultaneous moves by the two players as they interact with each other and with the rules of the game. How might we use these game rules to arrive at conclusions? We want to do something with them in a way that will help us evaluate or understand the system of which they are a part. We also want to use them to study each player individually.

Object System Defined

At this point we must introduce a special term: *object system.* The object system is the system we want to study; it is the "object" or subject matter of an investigation or learning experience. If we could study the object system directly we would not need a simulated system to experience or to experiment with. The object system is sometimes called the real world.

The object system of our example is all the real-world instances of real persons playing our simple (but expensive) game of conflict. In our example, the game rules are only some of the parts of the object system. The remaining parts of the object system are two real persons, each with $100,000 starting wealth. Of course, this object system may not now, or may never, exist, but such a condition does not obviate the usefulness of our example or even of simulation in general.

Models of Object Systems

A model is a representation of something else, as, for example, the model aircraft flown in a wind tunnel is a representation of the actual flight of an "object system" aircraft.

In order to construct a model, we need to know something about the object system we are interested in. The knowledge we use to construct a model may be generally accepted laws or principles about object systems like the particular one we want to study. In the absence of such well-founded knowledge, we may make tentative assertions about the object system and then proceed to build a model that reflects these hypothesized characteristics. Usually we try to make the structure of the model correspond in some degree to the structure of the object system.

We may model only a part of the object system, or we may model it all. For the object system of two real players playing our example game for high stakes, the rules of the game are only part of the object system. To model this system, we would incorporate these rules into our model, except perhaps for the high stakes. Thus the rules themselves (with adjusted payoffs and starting wealths) become a model of part of the object system. This illustrates the unaltered use in a model of some of the actual features or relationships of the object system itself. To model the whole example game object system, we would need to represent the two players in a way suitable for our purposes. Examples of both partial and complete models of the example game are given in Chapter 2.

We may use any convenient and useful device to construct our model: unrelated physical material, as in the case of the cisterns, tubes, and water used to simulate a market; scaled-down physical analogues as in the case of the model aircraft; an exact duplicate of part of the object system itself, as in the case of the space capsule used to simulate emergency conditions; combinations of actual parts and hypothesized behavior, as in the case of war games; word descriptions, as we shall use the rules of our example game; or mathematical formulae, as in the case of computer programs that simulate the economy.

Simulation Defined

Simulation is simply the dynamic execution or manipulation of a model of an object system for some purpose.

A simulation of an object system never attempts to become part of the object system itself. Thus this definition excludes the behavior of the tree frog that simulated its surroundings in order to appear to be part of them. It may even be impossible to try to make the simulation into the object system. For example, giving "subject" players of our example game each $100,000 free to start the game would not create that object system of interest to us wherein players voluntarily come together by some self-selection process and bring to a play of the game their own, but not free, starting wealths.

However, if the simulation becomes highly influential in guiding decisions that control the object system, a new simulation may be executed for a larger object system that includes both the old object system and its simulator. This is not unusual in the computer field where it is common for one computer to simulate another. If the second computer were simulating some object system, the first computer would be simulating a simulator. This practice could lead to an infinite progression of simulations. As a practical matter, for simulation work to proceed, the user must focus on a particular object system that remains within intended conceptual bounds during the period of the study.

Purposes of Simulation

What object system will be simulated depends on the specific purposes of the builder and user of the simulation. Purposes can be as varied and diverse as human activity. However, purposes may be classified in a general way as follows:

1. Simulations to aid our understanding of the workings of object systems.
2. Simulations to aid decision makers who control some aspects of object systems.
3. Simulations to train persons in the existing knowledge of object systems.

Where the use of simulations has become significant in controlling object systems, the three purposes become combined and interwoven. In some military defense simulations, the model that is dynamically executed is the concrete expression of the military strategists' understanding of their own defense system. Construction of this model forced its builders to a resolution of this understanding. Should the defense environment itself change, any new understanding would be tried out in experiments with the model. These experiments may lead to even further refinements in the state of knowledge of the defense system. By trying out alternative tactics on the simulation model, military planners can evaluate each alternative so that when the time comes to take decisive actions, they have some pre-emergency notions about the consequences. Where the defense system requires many employees who must be trained for their roles of vigilence, the simulation permits training exercises for situations that would be dire indeed if they were actually to occur.

Four Techniques for Studying Object Systems

We shall categorize four techniques for using models to study object systems. These techniques are not strictly separate from one

re Simulation

another, but they form a convenient way to talk about models and about simulation. One technique may lead to the use of another, or may serve as a basis for comparison of the results of another. All of them start with an initial model of all or part of the object system to be studied. The four categories of techniques are:

1. Analysis.
2. Man-model simulation.
3. Man-computer simulation.
4. All-computer simulation.

These four techniques will be illustrated in the next chapter by applying them to our example game.

EXERCISES

1.1. Think of decisions you have made in the past where you would have welcomed an opportunity to gain experience with the object system before committing yourself to action. Write a brief description of three of these decision situations.

1.2. Experiments, field trials, or pilot studies directly in the object system itself may be (a) impossible, (b) too expensive, or (c) too risky. Write a separate object system description to illustrate each of the above conditions.

1.3. With a friend, play the game of matching pennies described on pages 4 and 5 of this chapter (except do not include real money payoffs). For the first 50 trials, let your friend make his decisions by flipping his penny while you deliberately choose heads or tails. For the next 51 trials, let your friend try his skill at making reasoned deliberate choices while you make your own decisions by flipping your penny. Keep a record of the sequence of heads and tails each player chooses. Calculate the ending wealth position of each player. Write an explanation of the results. Save your results.

1.4. With a friend (perhaps the one from Exercise 3), play the game of matching pennies described on pages 4 and 5 of this chapter (except do not include real money payoffs). For all 101 trials, both you and your friend are to choose either a head or a tail by individual reasoning and deliberation. Do not relegate your decision-making responsibilities to flips of your penny. Keep a record of the sequence of heads and tails each player chooses. Calculate the ending wealth position of each. Write an explanation of the results. Save your results.

1.5. Three general purposes of simulation were given in this chapter. From your experience, describe a separate object system to illustrate the need for accomplishment of each purpose. Then use any of the three you have described, or a new object system if needed, to illustrate the need for concurrent accomplishment of all three general purposes.

2

FOUR TECHNIQUES ILLUSTRATED

The four techniques for studying object systems are:

1. Analysis.
2. Man-model simulation.
3. Man-computer simulation.
4. All-computer simulation.

Object systems were defined as the parts of the "real world" we want to study. As an illustration, an example game object system was described. In this game, two players, each starting with $100,000, match pennies for $1,000 a trial until one player is bankrupt or until they have made 101 trials.

The four techniques will now be illustrated by applying them to the study of the example game object system.

Analysis

One of the first things we could do to study our example object system would be to review our ideas about game players in general and then try to relate these ideas to the specific rules of the example game. Next, we would break these notions down into smaller elements. Then, based on experience or other information outside the present study, we would make assumptions about these elements and about relationships among them. Finally, we would synthesize these details and relationships into an overall conclusion about the behavior of the system. This technique is called *analysis* and it depends heavily on existing mathematical doctrine.

The "thought experiment" one goes through in performing analysis is in one sense a simulation of the behavior of the system being studied. However, many persons do not feel that mathematics looks much like any real system at all—indeed, Bertrand Russell, one of the world's most famous philosopher-mathematicians, has said as much (while simultaneously noting the remarkable usefulness of mathematics in dealing with real systems).[1] These persons do not like to give the name simulation to a study that is exclusively analytical mathematics because such manipulations do not "give the appearance of" the dynamic aspects of the object system studied. Of course, to be useful, analysis must give the effect of the object system to some degree.

So far we have stated that the game rules (adjusted for payoffs) are a model of only a part of the system we are interested in. The model can also be defined in such a way that the model represents the entire system. It is precisely this that mathematical analysis does. In order to make an analytical study of the whole game object system, assumptions about the behavior of each player must be made. Thus, analysis augments the game rules in order to create a mathematical model of the entire system. Unfortunately, by analysis there is no possibility of studying individual behavior of the players because specific reaction patterns must be assumed for them in order to arrive at analytical conclusions about the whole system. In other words, analysis discovers for us the implications of our assumptions, but it contributes nothing to the assumptions themselves.

Let us analyze our example game by assuming we know absolutely nothing about players in this kind of game. Thus, each player from our view is as likely to expose a head as he is to expose a tail on any trial. Our assumption means that the probability of a head

[1] Bertrand Russell, *Mysticism and Logic* (New York: W. W. Norton and Co., 1929), p. 75.

is one-half and that this probability does not change from trial to trial.

Since there are two players and since their actions are assumed to be independent of each other, each showing heads on his penny half of the time, we would expect to see heads on both pennies only a fourth of the time. This expectation is itself an analytical conclusion using the multiplication rule of probability theory. (Probability theory is not covered in this book.) Thus Player R would win on the average by matching heads one-fourth of the time. The same is true for matching tails. Altogether he would win on the average half the time and this implies that Player C would lose half the time, on the average. So Player R would win about as often as he would lose and we conclude by analysis—under the assumptions we made—that we should expect Player R neither to win nor to lose much money over a long number of trials. Player C's welfare is strictly determined by Player R's outcome, so neither player would be expected to change wealth significantly and the game would end after 101 trials according to the rules. But possibilities for variations exist. These are called chance variations when they are otherwise unexplained. Because of chance variations, we would be very surprised if our computed expectations actually came true in any one play of this game. It would be possible, but analytically a very rare event, for one player to win 100 consecutive times—thereby stopping the game because the other player ran out of funds.

(We shall use the method of analysis on the bodies of data generated during simulations. In data analysis, we attempt to break down the complex of data items into fundamental or meaningful classes and to discover relationships among these classes. We are distinguishing here between analyzing an object system and simulating one. However, no such distinction applies to studying data as data; we use similar methods of analysis of data whether the observations were made on real object systems or on simulated systems. Ultimately, all knowledge based on observations contains some portion of rational analysis of those observations.)

Man-Model Simulation

We started our simple example with the rules of the game. These rules did not provide enough assumptions to arrive at conclusions about the behavior of the entire object system of interest to us. In the method of analysis we expanded the original rules by adding assumptions about the behavior of the players. This was done to create a model of the whole object system. We then derived the implications of these assumptions.

Another way to represent the object system is to replace the assumptions about the players with real live players. The official action of each player in this game is to expose a head or a tail at each trial. We do not admonish the players to arrive at their decisions in any particular way. But because the players are aiding us in our study, we do not require that the loser pay $1,000 to the winner at each trial, and because we have a limited amount of money for research, we do not give the players $100,000 to consider their own so that we could enforce the original payoff rules. If we did require the $1,000 payments, if the players had paired off by voluntary self-selection, and if each had brought his own $100,000, we would no longer have a simulation but we would have an instance of the object system itself. Instead, let us set the payoff at one penny and give our live players an initial 100 pennies with the understanding they may keep whatever they have on hand at the end of the game.

This kind of simulation is essentially a laboratory experiment. The modified game rules form a model of part of the object system. This model is part of the experimental design. The live players are subjects in the experiment and they interact with each other and with the model, hence our term *man-model simulation.*

The results of such an experiment are data. *Data* are recorded observations. In our example, an *elementary observation* at each trial is noting the side exposed by Player R and the side exposed by Player C.

We could also record the total pennies on hand for each player at the end of each trial and the conditions under which the game terminated, but because the rules are so explicit this information can be derived from the elementary observations.

A more gross record of observations might report only whether Player R won or lost at each trial. This record would be simpler and would also let us derive the ending conditions and the total pennies for each player throughout the play of the game. However, it would not provide data about runs or sequences of heads and tails shown by each player, data that may be useful for understanding individual behavior in this experiment.

Let us imagine that we have conducted such an experiment with five pairs of subjects with the following results:

	No. Times Won		
Pair No.	R	C	*No. of Trials*
1	45	56	101
2	73	28	101
3	0	100	100
4	52	49	101
5	39	62	101

According to our original analysis we would expect winnings to be about equal, differing only by chance variations. Based on probability theory, the results for Pairs 1 and 4 are within the boundaries of expectation (where 90% of all possible results would fall).

However, under our previous analytical assumptions, the result for Pair 5 is rather rare. Number of wins exceeding 61 would occur only about 1% of the time in a large number of trials. The result for Pair 2 is even more rare. Number of wins exceeding 72 would occur less than one time in a thousand. To try to understand this observed behavior, we might want to question these players about their strategies and to search our observations for explanatory patterns of heads and tails.

The result for Pair 3 is so rare that perhaps we ought to suspect either a misunderstanding of the rules or a collusive agreement between the players. A comparison of the two sequences of heads and tails for the players of Pair 3 might be very informative. If Player R always showed heads and Player C always showed tails, we would probably feel confident about our suspicion of collusion.

These inquiries pertain to individual behavior. The performance of the entire system is also an important experimental result. For this experiment, the behavior of the whole system was rather balanced for Pairs 1 and 4, uneven for Pairs 2 and 5, and apparently extremely discriminatory or collusive for Pairs 3.

Our original interest was in an object system involving $1,000 payoffs. We have now learned something about the behavior possibilities for a simulated system that restricted payoffs to one cent. Does the result of our experiment tell us anything about the object system? It certainly tells us much about systems like the object system and the insights and understanding provided by the experiment may assist us in dealing with such object systems, but the results of this simulation study do not provide absolute conclusions for any particular instance of the object system itself. Only a direct experiment on the object system itself can do that. But then one reason we resort to simulation as a method of decision making and research is that such direct experimentation is impossible, too expensive, or too dangerous.

Man-Machine Simulation

Many representations of object systems are achieved through the use of machines. The space capsule simulator that solved the in-flight space emergency was a machine. The system of cisterns and tubes that represented a market was a machine. Airline and military pilots receive training in mechanical and electrical devices that simulate "blind flying" conditions. There are many situations in education where physical devices are used to simulate the real system because training in the actual situation would be premature, dangerous, expensive, or even impossible. These are also the reasons in scientific inquiry for using physical simulators for research in interaction with human subjects. A general name for this technique is *man-machine simulation.*

In man-machine simulation in training programs, students interact with some physical device. Given a responsive action by the student, the device feeds another stimulus back to him. In this way the student learns by discovering his errors and by being reinforced by his successes.

Man-machine devices usually provide controls over the kind of output fed back so that these machines are also often used to conduct experiments. One purpose of these experiments is to study the unknown behavior of subjects within a given and known simulated environment. Another purpose is to evaluate the effect of alternative total system configurations given known kinds of subjects employed.

Let us illustrate man-machine simulation by substituting a machine for one player in our man-model simulation of our example game. Let Player C be the experimenter. Of course, we want Player R to think that Player C is another player like himself since we are simulating an object system of legitimate players, not experimenters. Now let the experimenter, in his role as Player C, use a bent coin to make his decisions. Further let us assume this particular bent coin has been closely observed in the past with the results that in 10,000 tosses it has come up tails 6,537 times. This is very close to two-thirds of the time so we shall postulate that the probability of tails for this coin is two-thirds.

The coin is now a simple machine that can be operated by tossing it. The decision role of Player C in our proposed run of this experiment will be abrogated to this machine. When live Player R plays this game he interacts both with the model and with the machine that represents Player C. This is an example of man-machine simulation.

Man-Computer Simulation

In recent years, digital electronic computers have become versatile machines for man-machine simulation. They are versatile because through instructions they can be made to feed back output that gives the appearance of or the effect of a wide variety of object systems. When the machine is a digital electronic computer, we refer to the experiment as *man-computer simulation.* The only man-machine simulation we will be concerned with in this book is man-computer simulation.

There are three main classes of computers: mechanical, analog, and digital. Mechanical devices are rather restricted and are commonly called calculators. They represent decimal digits by positions of wheels and gears. Any device that provides a physical or electrical analogy that is converted into readable scales is an analog computer; examples are automatic speedometers, slide rules, flight simulators, and the water-level representation of a market economy mentioned in Chapter 1.

Digital computers manipulate numbers and symbols represented electronically by binary digits called *bits.* A bit is discrete while scales on analog computers are continuous. Bits can take on one of two states, usually symbolized as zero or one. Strung together, bits form numbers in the binary number system and can achieve high degrees of accuracy compared to reading from the scales of analog computers. Because of this accuracy and the capability of representing symbols in general with bits (e.g., the alphabet), we further restrict our interest in this book to man-computer simulation where the computer is an electronic digital computer.

To use a digital computer in place of the bent coin in our example, we must represent the tossing of this coin by computer instructions written in the form of a program. Once the computer is under control of the program, it can then write out on its typewriter, on its printer, or on a remote teleprinter or cathode ray tube, a sequence of the words "heads" and "tails" to give the effect of tossing the real bent coin. There is a formal theory about simulating random processes like the tossing of coins and the method used in the computer's program to select a "head" or "tail" at each trial is based on this theory. This is the method referred to earlier as Monte Carlo techniques, which we deal with in detail in a later chapter.

To perform our man-computer experiment we require one more control. The computer must be signaled when to "expose" its bent penny. One way to do this is to reserve an entire computer for the time of the experiment. Then each time Player R is ready, the experimenter presses console buttons or typewriter keys to activate the computer's program. Another way is to program these signals into the original set of instructions so that the computer asks when Player R is ready. Player R or the experimenter then presses a key and the computer prints or displays a "head" or a "tail" and again asks when Player R is ready. This method is especially appropriate when the experimenter does not control the entire computer but is conducting his experiment at a station remote from the computer such as a teletypewriter or a keyboard with a visual display. The device at a remote station is called a *remote terminal.*

Let us imagine we shall conduct a man-computer version of the example game as an experiment. The computer will be programmed to simulate our bent penny which has a probability of tails equal to two-thirds. Let us further imagine we have three subjects who have access to remote terminals. We could run the experiment for all three subjects simultaneously if we have three remote terminals, or sequentially if we have only one remote terminal. Under our original analytical assumption of ignorance about the players, we hypothesized that Player R was as likely to show a head as a tail. In this case, regardless of the side of its penny "chosen" by the computer, Player R's chances of matching it remain one-half, and we would expect most subjects to win about as many times as they lose. Imagine that we ran the experiment and the following results were obtained:

Subject No.	*No. Times Won*	*No. of Trials*
1	48	101
2	63	101
3	81	101

Of these results, only Subject 1 behaved within the boundaries of expectation given by our analysis. The results for Subject 2 and 3 are so rare, under our analysis, that we should look for some explanation, both by questioning the subjects and by inspecting the sequences of heads and tails.

We might find on examination of the data that after only a few trials, Subject 2 always chose tails. In reply to our questions, he might state he discovered that tails were occurring more frequently so he maximized his winnings by always choosing tails.

Inspection of Subject 3's decisions may offer no clue to his success, which is extremely rare when viewed by our original analysis. If we question him he might reply that he too noticed the bias of the computer for tails but that he was soon bored showing only tails trial after trial. When he noticed a pattern evolving in the computer's choices he tried to match the computer event for event and happily surprised himself with his own success.

At this point we would want to question Subject 3 further about the pattern he thought he had uncovered, and then we would want to see if he were indeed correct. The pattern he believed he found may have been the chance result for these particular trials of the computer's bent coin tossing simulator and would not occur again for hundreds of thousands of tosses. On the other hand, we might find this pattern recurring every few trials. If so, we might want to investigate and revise our computer program. Or we might want to use the program as a pattern generator for man-computer research into human pattern recognition.

Man-computer simulation is a powerful educational, scientific, and managerial tool. It can be used for training students in simulated environments, for research into human behavior under controlled conditions, and for investigating alternative system policies using as subjects the actual persons who work in the object system.

All-Computer Simulation

Digital electronic computers have evolved to the point where their influence on society is called a revolution. They perform computations and manipulate data and symbols with almost unbelievable speed. By taking over repetitive chores that hitherto consumed human mental effort, they expand the potential of the mind of man to dimensions that presently appear without bounds.

The impact of computers has been threefold:

1. They calculate answers to mathematical problems fast enough for the results to be useful. Answers now obtained in minutes heretofore required years of human effort. Before computers, such calculations would not be attempted because by the time the answers were found the problem would be obsolete. Without computers space explorations would not exist because trajectory corrections for space vehicles require immediate computation as the rocket proceeds in its brief powered flight after launching.
2. Computers rapidly process masses of data. Periodic reports for business and government are now delivered to decision makers in time to be relevant. Information can be retrieved from large bodies of electronically stored data almost instantly. Procedures in our interdependent society such as clearing checks in banks or reserving seats on airliners are accelerated by computers. Computers also speed up communications and production in other ways in modern society by supervising message handling and by monitoring on-going processes.
3. Computers are logic machines and general symbol manipulators. As such they literally extend the mind of man in man's own fashion. A human problem solver using a computer need think through a general problem only once, provided he articulates his thoughts in a language computers can translate into instructions. He can then use this computer program again and again to "think through" the same problem, but for different data or for different circumstances.

Studying object systems by computer simulation requires all three of the above properties of computers. Computation and data processing are necessary, but the key feature of computers for simulation is their role as logic machines and symbol manipulators. The latter property enables an experimenter to conduct experiments "in the computer." With appropriate programming, the flashing of electronic signals through the intricate circuitry within a computer becomes a simulated experiment. The resulting data may be studied in the same way as the data from an experiment in a conventional laboratory or with live subjects.

We have already considered experiments in which live subjects interact with a computer that simulates part of an object system. This was called man-computer simulation. When the entire experiment is conducted by the computer as it executes its program of instructions, we shall call the experiment *all-computer simulation*, or frequently just computer simulation.

In all-computer simulation, experimental controls are provided through the program of instructions. The program also provides step-by-step representations of the behavior we want to study. The experimenter can rerun his experiment by changing the experimental controls while using the same simulated behavior, or by changing the simulated behavior while using the same controls, or by changing both controls and behavior.

An all-computer simulation is merely an electronically assisted "thought experiment." No features can ever be included in a computer simulation that have not first been thought through into the form of computer instructions by the experimenter and his staff. The power of all-computer simulation comes from the speed and scope of the execution of the computer program and hence from the great flexibility available to man to conduct alternative highly complex thought experiments. The number of trials investigated in computer simulated experiments is many orders of magnitude beyond hand computation. Computer simulation can also combine into a single experiment thousands and thousands of interacting parts and facets that otherwise might be manipulable only in the real object system, where experimentation may be impossible, too dangerous, or too expensive.

The practical power of all-computer simulation arises then from these two sources: 1) the ability to replicate experiments rapidly, and 2) the ability to combine enormous numbers of subsystems into a single experiment. Thus, computer simulation extends the mind of man by hastening the speed of thinking and by enlarging the scope of what can be thought about in a single context.

Note that we have not restricted computer simulation to any particular aspect such as mathematical computation, data processing, or logic. As general symbol manipulators, computers appear able to simulate object systems expressed in almost unlimited ways, provided someone has created a computer language that the problem can be stated in.

Let us run our example game in an imaginary all-computer simulation. This time we shall simulate players observing the $1,000 payoff rule. First we create a computer program that is a model of the original game rules. Then we write computer programs to represent Player R and Player C. We also write programs to record and process the data such as the sequences of heads and tails generated by the experiment and to calculate during the simulation run various statistics such as each player's current proportion of wins.

Also, to illustrate the flexibility of computer simulation, let us manipulate our assumption about the starting wealth of each player. After all, we can start the players in the computer simulated game with any specific total wealth. The amount for each player need not be the same. Say we "give" $30,000 to Player R and $17,000 to Player C. Let the game end when either player has zero wealth, or after 101 trials as before. The controlled conditions for our all-computer experiment now include the rules of the game for each trial, the amount and distribution of the starting wealth, the payoff amount, and the maximum number of trials. If we change these conditions we change the experiment, which we may want to do, for example, in order to study the effect of various wealth positions on total system behavior. To simulate that part of the object system that is Player C, let us use our previous man-computer simulation program. This model of Player C's behavior will choose tails with probability two-thirds, but with no way to predict exactly the next decision from its prior sequence.

To simulate Player R in the computer, let us stipulate that he will always show tails. By analysis we would conclude that Player R would win about two-thirds of the time and on the average win in the neighborhood of $67,000 in 101 trials. But in the experiment we will run, Player C would run out of money long before this happens. On the other hand, Player C could win all of Player R's $30,000, but this would be analytically extremely rare.

If we change the simulated behavior of either player, we change the experiment. Such a change is analogous to the change in a live experiment resulting from using a different kind of subject.

Assume we run the simulation once with the result that Player R has won all of Player C's wealth on the 79th trial. What does this result mean? It certainly does not mean that the object system would always end on the 79th trial, even if we have simulated it perfectly. It means, presuming valid simulation, that ending on the 79th trial (and Player R taking all wealth) is one of the many possibilities that could occur in the object system.

Therefore, let us run the same experiment again. This time it ends on trial 56, Player R the winner of all wealth; this also is one of the many possibilities in the object system. We try again and this time the play ends on trial 65 with Player C taking all. In this last run, it took Player C 65 trials to take $30,000 away from Player R, while Player R in the two previous runs of the experiment required 56 trials and 79 trials to take only $17,000 away from Player C. While we may view Player C as "just lucky" in this last run, we must recognize that this result too is a possibility in the object system.

We could replicate this experiment until we had generated a large number of results. These results would then provide a distribution of the possible object system performances and we could use this distribution to arrive at conclusions for dealing with object systems like our game. For example, we may conclude that the strategy of mixing decisions randomly but in uneven proportions may be frequently ruinous to a player starting with only $17,000.

Next, we could continue our investigations of our example game model by "experimenting" with our experiment. We would do this by programming different decision strategies for the players, or by starting the players with different wealth positions, or by changing both features at once. For each change, we would run a set of replications of the new experiment on the computer and study the resulting data for implications about the object system and also for hints on how to improve our simulation of the object system.

Simulation entirely in the computer is completely conceptual. Its usefulness depends on the applicability of our concepts to the world about us. Computer simulation is the processing of concepts to discover their consequences. The results are themselves only conceptual. But this is the essence of the decision process in our society. A decision is only conceptual until an attempt is made to implement it. Computer simulation is an aid to decision making, whether the decisions concern games, scientific inquiry, public policy problems, or evaluation of alternative business strategies.

EXERCISES

2.1. Describe your play of the example game in Exercise 3 of Chapter 1 as a man-machine simulation. Distinguish between the simulation and the object system. State the roles of the players, the model, and the machine.

2.2. Describe your play of the example game in Exercise 4 of Chapter 1 as a man-model simulation. Distinguish between the simulation and the object system. What constitutes the model?

2.3. Review your explanation of the results of your playing the example game in Exercises 3 and 4 of Chapter 1 in terms of the technique of analysis. What is the object system? What constitutes the model?

2.4. Repeat Exercise 3 and 4 of Chapter 1. Explain any differences in the results.

2.5. List the logical steps needed to think through as experimenter a run of Exercise 3 of Chapter 1. Your list should form a "program" of steps that you as experimenter would have to make sure were accomplished if two other persons were the players. Could another experimenter execute this program without having to ask any further questions of you?

2.6. List the logical steps needed to think through as experimenter a run of Exercise 4 of Chapter 1. Your list should form a "program" of steps that you as experimenter would have to make sure were accomplished if two other persons were the players. Could another experimenter execute this program without having to ask any further questions of you?

2.7. Since, as in the man-computer example in this chapter, electronic computers can play the role of live decision-making persons, it could be claimed that computers can think. Write an argument either for or against this claim.

3

MODELS FOR SIMULATION

In this chapter we shall examine the role of models in simulation. To do this we shall first consider the role of models in general and the relation of models to decision making. This will require a brief consideration of man's knowledge and how he uses this knowledge in interaction with himself and with the world around him. The practical application of simulation to problems in government, industry, education, and science depends on these foundations.

Theory

Man accumulates knowledge over time. This is done by passing general or abstract principles on to next generations. But rarely do the general principles remain the same. Man changes his general principles whenever by doing so he can better understand himself, his world, and how to live in his environment.

A theory is a set of these general or abstract principles as they exist at some moment in time. Theories for particular subject areas are usually interrelated sets of principles confined to manageable domains. We have, for example, the economic theory of the firm, the theory of games, nuclear theory, learning theory, role theory, and hundreds of others covering the whole of man's knowledge.

When we begin an inquiry, or when we start working toward a solution to a problem, we do not come to the situation with blank minds. We start with what we know from existing theory for that subject area, otherwise our initial investigations would founder in a mass of confusion. In fact, confusion implies lack of discrimination and hence a lack of understanding and the absence of knowledge; that is, confusion means the lack of a theory.

Hence, it is the theory at hand at the outset of an inquiry, or when we attack a problem, that enables us to start at all. Yet before we are finished with our activities, our findings may very well change the existing theory. Alternatively, they may confirm it.

We constructed our example game from existing knowledge of games. The conclusions we drew by the method of analysis were drawn from the existing theory of games. Our imaginary simulations, however, revealed behavior in games that varied significantly from the theory of games. Game theory is relatively new and if these results were consistently found in actual replications of these experiments, they would contribute to the ongoing evolution of our knowledge of human behavior in games and game-like situations. In fact, our imagined examples are simplifications of experiments that have already made such contributions.

Theories are purposely broad; they try to generalize over many specific cases. This is really an economy of communication so that man does not need to pass hundreds and hundreds of detailed instances from one generation to the next. An object system is one of these detailed instances. To use theory, one must make the generalizations of theory specific enough to guide the making of observations and the taking of actions. Models serve this purpose.

Where our knowledge is weak, which means we do not have a good theory at hand, we may tentatively need to assume some characteristics of object systems in order to proceed. Tentative statements from which we proceed are called hypotheses. (The word hypothesis is made up of the prefix *hypo*, which means less than, slightly, or somewhat, and *thesis*, which means a proposition to be maintained under debate or against objections. Thus, a hypothesis has not yet attained sufficient status to become a "good" thesis, that is, a part of a good theory.) To use or test hypotheses, we must also make their general propositions specific enough to guide the making of observations and the taking of actions. Again, models serve this purpose.

Model Defined

A *model is a constructed specific expression of a theory or of one or more hypotheses.* The model aircraft "flown" by engineers in a wind tunnel is an expression of the theory of flight. A full-scale flying aircraft patterned after the model is an actual application of theory. Its flight is an instance of the real thing the theory is about, i.e., the object system the model represents.

Any given theory may lead to almost unlimited numbers of different models and applications. Frequently actual applications are preceded by several models that have been used to explore many alternative features before a final design is chosen. This again illustrates the fact that a model need not represent something that exists. Many models represent hypothetical designs that are future events that never occur.

Road maps are also examples of models. They are specific expressions of our knowledge of the surface of the earth. They do not express our whole knowledge, only that required for motoring. Theories and models are not infallible, as in the case of the road map that is wrong because a road is closed or a bridge washed out. Old road maps that do not show new routes illustrate the need to review our theories and their specific expressions (models) as time passes.

Returning to the definition of a model, by "constructed" we mean that models do not appear naturally—they are the creation of man. We create models to serve our purpose. Who are *we*? In this book we have in mind educators, researchers, and managers who are the decision makers in science, industry, and government.

What we mean by "specific" will be explained later. However, here we repeat that it is our initial existing theory that provides the framework for specification in model construction. Without a theory, we have no starting point for building a model. Even a model expressing a hypothesis is founded on existing theory but with hypothetical variation.

Most theories are general, in fact too general to be tested in any meaningful way in their entirety. General theories are tested only through specific expressions of them, that is, through models designed to give operational opportunity to the implications of the theory.

To test a theory or hypothesis, first a model is constructed. Then the model is used in an application, called a pilot study or a field test, or in a laboratory experiment. The results of the test or experiment either confirm the theory, disprove it, or lead to further tests that may modify the theory. The relation between a theory and a model is illustrated in Figure 3.1.

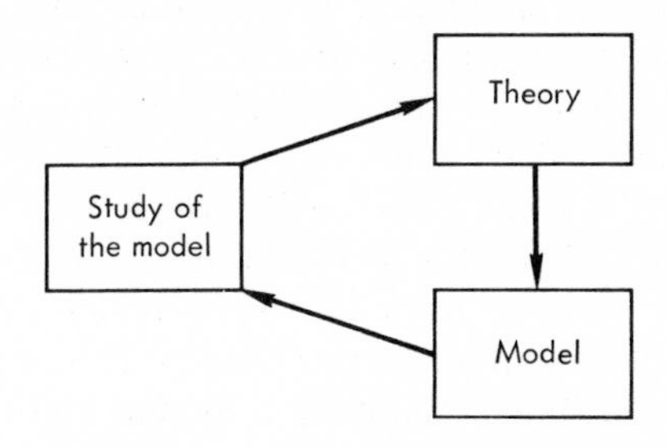

FIGURE 3.1. Theories and Models

Notice in Figure 3.1 the circular path of the arrows connecting theory, model, and study of the model. Once the path returns to theory, another model may be constructed and studied and the cycle repeated. This is the process by which our knowledge grows. Such cycling rarely stops for long and new knowledge is almost always evolving.

Simulation Models

A simulation model has the following properties:

1. It is intended to represent all or part of an object system.
2. It can be executed or manipulated.
3. Time or a count of repetitions is one of its variables.

4. Its purpose is to aid understanding of the object system, which means one or more of the following:
   a. It is a (partial) description of the object system.
   b. Its use attempts to explain past behavior of the object system.
   c. Its use attempts to predict future behavior of the object system.
   d. Its use attempts to teach the existing theory by which the object system can be understood.

Of course, simulation models have all the characteristics of models in general: they are constructed as specific expressions of theories or hypotheses. The relations among theories, object systems, and simulation models are shown in Figure 3.2. If we replace in Figure 3.2 "experiments with the model" by "experience the model," we reflect the use of simulation in education and training. If we replace "experiments with the model" by "analyze the model," Figure 3.2 would apply as well to analytical models that represent object systems. We shall continue to be interested in analytical models in this book because they will form a basis for creating and testing hypotheses about the behavior of simulation models. We will not further concern ourselves with models that are not intended to be at least partially representative of some object system. For example, we exclude as simulation models mathematical formu-

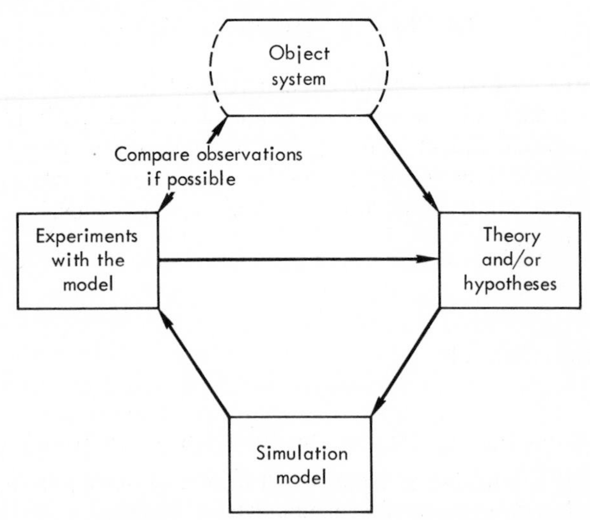

FIGURE 3.2. Theories, Object Systems, and Simulation Models

lations not referenced in some way to real phenomena. However, we should recognize that mathematicians regularly call many of their purely abstract creations mathematical models.

Both analytical models and simulation models can be manipulated, but analytical models by definition are not experiments. Study of the model to reach conclusions about the object system can be attempted by analysis, by experiment, or by both methods. Many times analysis alone will serve our purposes, and analysis is usually easier, quicker, and less expensive than an experiment. At other times, analysis may yield incomplete results or the model may be too complex for direct mathematical attack; in these cases, experiments are the only way we can use our model to reach conclusions about the object system it represents.

Simulation models are dynamic models and always involve change in the state of the model through time. By *state of the model* we mean the values existing for the variables at any moment in time.

Two terms about time need definition. *Real time* is the time we live in; it is ordinary clock time; it is the passing of real hours, days, months, and years. Object systems behave in real time. *Simulated time* is a variable in simulation models meant to be representative of real time. It is not necessary that simulated time bear a constant ratio to real time. For example, we may simulate a year by making certain comparisons and calculations prescribed by the model. This task if done by hand may at first take half an hour. Later, it may take only ten minutes. Yet each cycle through the model represents a year of real time. On the other hand, it may be convenient to tie simulated time to real time. For example, we could let an hour on a computer's real-time clock represent a real year. We might then program the computer to scan data received from remote terminals every real hour. Such inputs to the model would represent simulated annual changes in certain variables. In other applications, simulated time may pass moment for moment with real time, or even more slowly than real time.

The purpose of simulation models is to aid our understanding of the object system we are interested in. The use of simulation in education helps students learn how others understand the object system. Understanding means we can describe the object system to each other and we can explain what is happening in the system. The ultimate test of understanding is our ability to predict future behavior of the object system. If we can predict its behavior, we may then be able to control the object system in ways to meet our goals. Or we may only be able to understand why we cannot control it as we wish. Wherever possible the observations made on the model should be compared with observations on the real system in order to evaluate and improve our theorizing and model building capabilities.

Frequently the only description of an object system we have is the model we construct of it. Models as descriptions are almost always partial or incomplete because we select for attention only those aspects of the object system that appear related to the ultimate purposes that originally inspired the model building efforts. The broken line around "object system" in Figure 3.2 reflects our inability to capture "all" of an object system in a description; there is always something left out.

Occasionally, the model itself becomes the only theory for its subject area. In this case, since the theory is the model, there is no need to express it more specifically in order to use it. Let us carry this situation one step further and assume the model itself is written in one of the computer languages. Then, in this special case, the computer program is our theory of the object system.

The Ingredients of Simulation Models

Simulation models are simply sequences of operations. Sometimes these operations are continuous, as is the passage of air over an aircraft model. Another example of a continuous operation is the background noise in a man-model or man-computer simulation when this noise is a variable in the model. However, in this book we will be more concerned with operations that are discrete and that can be thought of as distinct steps to be executed or performed. Examples of such operations are: add one number to another; divide one number by another and replace the first one by the quotient; tally; count; repeat identical sequences of operations again and again; test a number or symbol; make comparisons; skip certain operations in the sequence depending on the test results; manipulate special devices; report results; move or place a token or counter on a card or board; emit sounds; actuate special displays.

When a simulation model is used, which means that operations such as those cited above are performed according to the prescriptive sequence of the model, there must be some input external to the model that determines the starting conditions for the operations. In addition, the operations themselves may be externally alterable to some degree.

Again we need some definitions. First, we define a *run of a simulation* as cycling through the operations of the model for a measurable amount of simulated time. Simulated time itself may be measured in terms of these cycles; for example, the sequence of trials in our example game is a sequence of "times." In some applications, when it is known in advance that nothing happens or changes during a cycle, the cycle may be skipped with the effect that simulated time progresses in uneven leaps.

Next, we define a *parameter* of the model as a number or symbol that remains constant during one run of the simulation, but that can be changed from run to run. For this reason, parameters are sometimes called variable constants. In contrast to input variables defined next, parameters are input constants.

Finally, we define a *variable* as an entity that can take on different values or be represented by different symbols during a single simulation run. Variables have names and we frequently use the name as if it were one of the possible values of the variable; the familiar n and x of algebra are examples. Variables are classified in this book into *input variables* (those arising external to the model) and *generated variables* (those arising as a consequence of the operations of the model). Input and generated variables are sometimes called *exogenous* and *endogenous* variables respectively.

Usually we are interested in only a selected few of the generated variables resulting from a run. A record of a single variable, either input or generated, showing its value at each moment of time during a simulation run is called its *time path. Starting conditions* are initial values given to input and generated variables. The first value in any complete time path is always a starting condition.

Outputs are the data we wish to obtain from a run of the simulation. These data may be records of parameters and starting conditions, time paths of input and generated variables, selected variable values from selected states of the model, or just the ending conditions. We may also create additional outputs that are combinations of or measures on the foregoing data.

Specificity in Simulation Models

Our definition of a model asserted that a model was a constructed specific expression of a theory or hypothesis. We have delayed discussing "how specific" a model must be until now so that these deliberations could apply particularly to simulation models. Simulation itself is the process of operating the model. Without parameters (if required) a model is not specific enough to perform its operations. Without input variables (if required) the operations of a model have nothing to "operate on." With a given set of parameters the model may operate in many simulation runs on different sets of values for the input variables. On the other hand, several simulation runs may be conducted on the same set of input variables but with different parameters. In other runs, both parameters and input variables may be changed. All three combinations would produce different sets of generated variables. Data selected from these generated variables are the output we want to study and compare as simulation runs are completed.

We see that we have a hierarchy of specificity. There are three levels: 1) the operations prescribed by the model, 2) the parameters that may change from run to run but remain constant during any one simulation run, and 3) the input variables that may take on different values at different moments of time during a run. A change at any of these three levels results in a change in the values produced for the generated variables.

The answer to the question "how specific must a model be?" can be answered in terms of its operations, parameters, and input variables. A model is specified by a unique set of operations. Any change in operations is a change in the model and creates a different, although perhaps equally appropriate, expression of the larger theory. "Experimenting" with the model itself by changing its operations is a valuable technique for building models. A change in a parameter for a run, or a change in input values, is not a change in the model—of course, such changes are changes in applications of the model and it is this flexibility that enables us to perform many experiments with a single model.

How specific must a simulation model be? The answer is: given its parameters and input variables, a model must be specific enough to run. This means that the separate operations must each be specific enough to be carried out by a person or by a computer or other machine. It also means one operation does not create ambiguity or unplanned inoperability for another operation. Frequently, achieving an operating simulation model is a rather demanding experience.

Levels of Specificity Illustrated

Let us illustrate these levels of specificity with the all-computer simulation of the example game from Chapter 1.

To do this we shall use a form of communication called a flow diagram.[1] A *flow diagram* is a graphic representation of the operations prescribed by a model, of the conditions under which operations are performed, and of the sequence of performing them. Flow diagrams also show operations external to the model such as the briefing of live subjects and the assignment of values to parameters and input variables.

A flow diagram of the all-computer model of the example game from Chapter 1 is shown in Figure 3.3. Input or output to a simulation model is shown by a trapezoid. Operations are shown by rectangles, except where diamonds are used to show a test operation that determines the direction of flow. Ovals represent start and stop operations, interruptions, or connecting points of model segments that are subdivisions of a larger or main model. Arrows indicate the sequence of operations, hence they show the path or flow through the model and through the larger system containing the model. A model is completely expressed by the symbols for its operations, but the system containing the model is

[1] As used in this book, flow diagrams are more general than the system and program flowcharts used in work with computers. Flow diagrams permit naming operations rather than showing them in complete detail, and they also provide nodes for decisions by live participants or by the experimenter. The author has tried as much as possible to follow conventional usage.

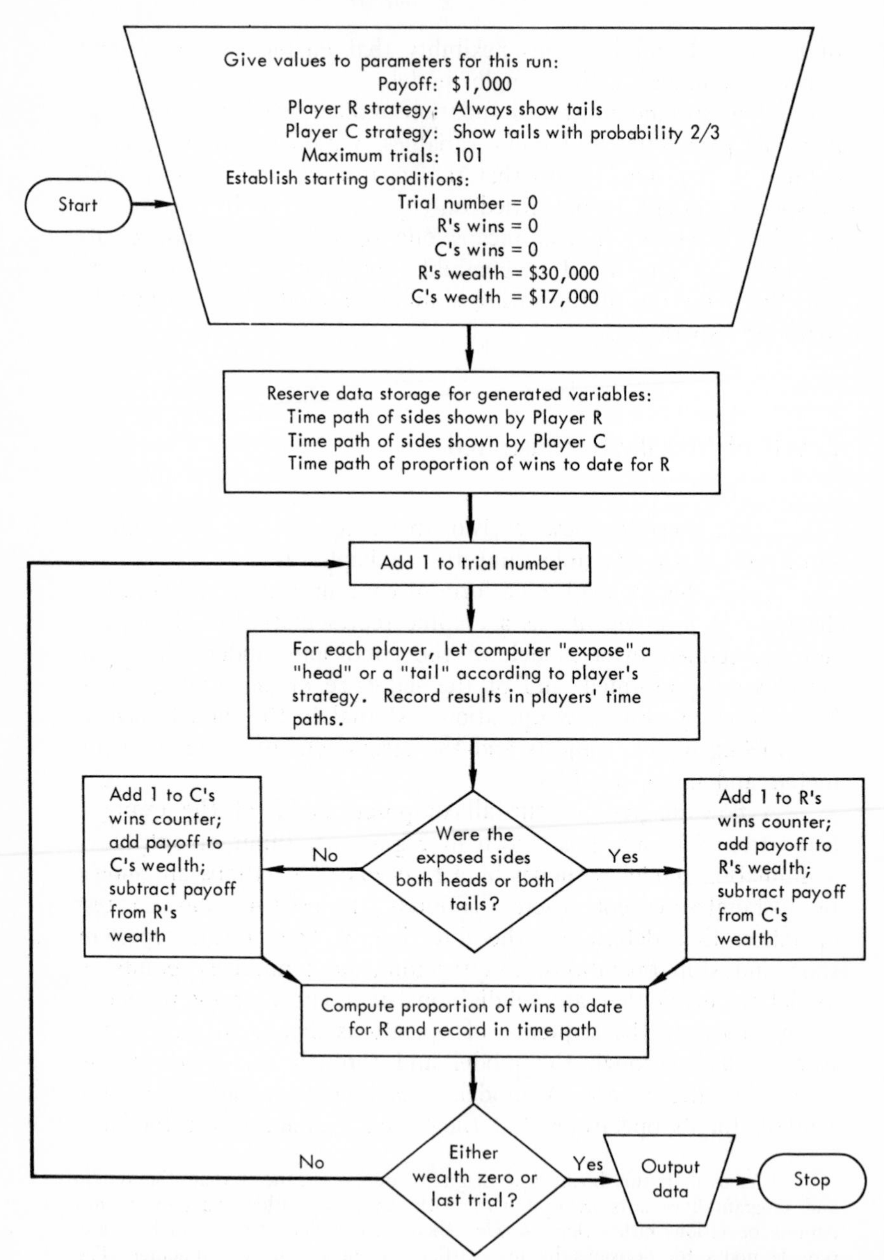

FIGURE 3.3. Flow Diagram of All-Computer Example Game Simulation

still not specific enough to run as a simulation until parameters and starting conditions are provided. The parameters in this example, which we can change from one run to the next, are the payoff amount, the players' strategies, and thee maximum number of trials. The starting conditions (in this case required for only certain generated variables) are the beginning wealth positions of each player and the initialization of the wins and trial number counters. It is conceivable that we may at some time want to study this system from some prior stage of play so that initialization of counters does not always set them to zero. There is only one input variable and it is hidden in this flow diagram; this is the input required by the computer program segment that "exposes" heads or tails for each player. The time path of values for this input variable is a set of random numbers. We implicitly assumed in Chapter 1 a different time path of this variable for each run of the simulation. How random numbers are created and used will be explained in Chapter 8 on Monte Carlo techniques.

The reservation of data storage shown in Figure 3.3 is of concern to simulation system designers whether the model is to be processed by hand or to be programmed for computer processing. In a flow diagram, data storage specifications indicate the generated variables we want to keep time-path records of. Notice in Figure 3.3 that all the generated variables receiving starting condition values will have changing values as the run progresses, and that we lose each past value as the new value is generated. This is not true for the generated variable called proportion of wins to date for R because we are keeping a time path of its values.

At the end of the run, we shall output from data storage the time path of proportions of wins to date and the ending values only for those generated variables not recorded in time paths, namely, trial number, R's wins, C's wins, R's wealth, and C's wealth. This output constitutes the results of the simulation run. For convenience in studying the results, we might also output a record of the parameters, starting conditions, and input variables.

In Chapter 1 we ran the simulation of Figure 3.3 three times, obtaining different results each time. The only difference between the runs was the time path of the input variable used by the "coin tossing" segment. This input variable was a "random" number. A new number was used for each player at each trial. These time paths of numbers were different for each run. If we had used the same list of random numbers each run, the results would have been identical for all three runs.

Simulation models are part of *whole simulation systems* (see discussions in Chapters 4 and 5). They are that part that is intended to represent all or part of an object system. The balance of the whole simulation system contains all the work and plans needed to use the model for the purpose for which it was designed.

Design of Simulation Experiments

Experiments performed with an all-computer simulation are a researcher's dream come true. Everything is controlled as he wishes. There are no effects outside his control. Of course, he may not understand everything that is happening in his simulation, but to arrive at understanding is one of the purposes of simulation.

Since everything is controlled in all-computer simulation, two or more experimental runs may be compared directly, without necessary reference to some other "control" run or to an analytical expectation.

In our all-computer example game, we ran the simulation three times, changing only the input time path of random numbers. The output of these three runs represents three of the many possible results had Player C really been using a bent coin with a probability of tails of two-thirds.

All-computer simulation gives us an opportunity the actual coin does not. We can run our experiment again using identical sequences of all "random" influences, a replication that is most unlikely with a real coin. If we change nothing at all, the results would always be the same. This provides no information other than that the computer is consistent. However, if we change the probability of tails of Player C's coin, we may not obtain identical results. In this way we can learn the precise effect of changes in the probability of tails for C's coin, because everything else, including the "random" influences, is held constant.

What other changes might we make for further experiments? We could hold the parameters and all but one starting condition constant, and change, say, C's wealth. Or we could hold the starting conditions and all parameters constant except one, and change, say,

R's strategy. Or we could change only the time path of the "random" input variable and obtain another three instances of the many possible for the original experiment. If we change more than one parameter or input variable at once, we may be unable to state how each change affects the new results, but we can observe the joint effects. Seldom can such ideal experimental control be achieved elsewhere. Experimentation with all-computer simulation models achieves the idealization of the perfectly controlled experiment.

When live subjects are introduced into an experiment, the idealized control of all-computer simulation disappears. Inevitably, uncontrolled variations are brought in by both the subjects and by the experimenter's handling of them. The effects of these uncontrolled variations are mixed in with the effects caused by changes in parameters, starting conditions, or input in variables. Here the experimenter should design the simulation run in accordance with the dictates of his scholarly discipline; usually this involves a control group and statistical considerations.

The essential feature of experiments conducted under simulated conditions is the capability of answering the question, "What would happen if such-and-so change were made?" Many "what if" questions involve changing the model. This is extremely important when the "what if" question cannot be answered by inspection, analysis, or experts, and the system changes cannot be tried directly in the real object system for reasons of morality, expense, or impossibility. Other "what if" questions involve changes in parameters, starting conditions, and input variables.

Thus, there are three levels of controls for experimenters, just as there are three levels of specificity. Experimental changes can be made in the model (change its operations), in the parameters (change the input constants from run to run), and in the input variables (from run to run, changes in starting conditions, sequences of input values, or live subjects).

Design of Educational Simulations

When live subjects interact with the simulation model, they learn the model, and, hence, portions of the theory it expresses. The controls that enable experimental variations also enable adapting a simulation to particular students. (See the discussion of *verisimilitude* in Chapter 4.) Parameters and starting conditions may be changed from one run to the next so that students do not "pass the word" on how to "beat" the simulation exercise. Input variables can give students experience with trends, seasonal fluctuations, or unexpected phenomena. Monte Carlo model segments can give students experience with uncertainty, an experience that is in distinct and sometimes uncomfortable contrast to the determinism of classroom examples and case histories. The simulation can be stopped and rerun from any point so that students may see "what if" they had done something else.

Purposes of Simulation Designs and Execution

A simulation model, as defined earlier in this chapter, was required to serve one or more of the following purposes:

1. Describe the object system.
2. Explain past object system behavior.
3. Predict future object system behavior.
4. Teach existing object system theory.

An essential ingredient of the practical and successful design and use of simulation models is a clear focus on purpose. At each stage of model development, the relation of what is being done to the ultimate purposes of the project should be reviewed. This requires that at each stage the project purposes be specific enough so that this relationship may be assessed. But this does not mean that the ultimate purposes may not be changed from stage to stage. Simulation is as effective a way to examine purposes as it is to explore an object system.

Practical matters of the design of simulations are the topics of Chapters 4, 5, 6, 7, and 8. While ultimate purposes are not frequently reviewed in these chapters, the reader should pause from time to time to reflect from the specific details being discussed to the ultimate ends they might serve.

Transfer to Object System (Validity)

Simulations representing object systems are run in educational applications to provide learning experiences in the theory of an existing object system. They are run in research applications to describe, explain, and predict object system behavior. Of course the simulations themselves are not the real thing. This requires us to ask the question: Will what is learned or what is best in a simulation work or apply in the real object system?

The answer to this question is difficult. (See, for example, Exercise 3.3.) The question could and should be asked of analysis, advice of experts, practical experience, or any other means for choosing actions in any facet of life. The plain fact is that regardless of method or technique this question cannot be answered when the answer is needed so much—before proposed action is taken.

Any decision when implemented impinges upon and changes the future state of reality from what it would be if another course of action were pursued. Truth in decision making lies in the future and cannot be known at the time of action. It is a matter of confidence if one believes what one is about to do will produce the desired results. For many object systems, such confidence is high. These systems are usually characterized by a high degree of regularity in their behavior. For others, confidence is low, either because there are few repetitive factors or because we do not know much about these systems.

A simulation model is partly an expression of what is known about an object system. Its use teaches what is known or finds the implications of proposed changes. If we have high confidence in what we already know about an object system, we can have high confidence in the results of our simulation study or in the effect of training by simulation. If our prior confidence is sufficiently high over a broad enough domain, we may not need a simulation study at all, even if analysis and direct experimentation are impossible. If we have low confidence, our efforts to build a simulation model may show us where first we should try to increase our knowledge. One justification often given for model building activities is that in order to get his model to run, a decision maker is required to be very specific about his problem and about his alternative approaches to it, and that forcing a person with a problem to be this specific is meritorious in itself. This, however, is only a side benefit of simulation model building. The direct benefit is evaluation of alternatives, instruction in theory, and focusing on areas for further inquiry in the ongoing evolution of man's knowledge. As this knowledge expands, so does the transferability of simulation results to real object systems.

EXERCISES

3.1 Obtain a kitchen recipe for baking a cake or making a pie. Is the recipe a model? What part compares to the rules of the example game? Draw a flow diagram for baking n cakes or making n pies, where n is a parameter. Carefully distinguish between input variables and generated variables. Provide for appropriate time path records to reflect the step-by-step status of your n cakes or pies. If you executed the recipe, would this be a simulation?

3.2 Assume that medical scientists have constructed a computer model of a patient with heart disease. By changing parameters, this simulated patient can be given different sex, age, and health history. Input variables to this model are specifications for medical care and treatment into the indefinite future. Generated variables are the physical capabilities and health of the "patient." Of course, the simulation run ends when the "patient" dies, hopefully at a ripe old age. Imagine yourself a real live patient with a heart condition. Write your reaction to the idea that your doctor chooses his plan of care and treatment for you according to whether this same plan is able to "cure" the simulated patient in the computer.

3.3 The question of validating a simulation is sometimes agonizing for researchers and educators. To validate a simulation model means to show that it is true. This ideal condition cannot hold because all models are abstractions to some degree, and the only true model is the object system itself. Yet many simulation users claim that their models truthfully represent object systems.

In the chapter validity meant usefulness in the future, linked with the process of ever-evolving more useful general theories. Others have suggested that the question of whether a model is true of the intended object system or not is irrelevant. What they say is important is how two alternatives compare to each other and not what the "true" results would be if either were implemented. For this purpose, the model itself need not be valid or "true," but merely be an environment in which the alternatives tested can show their merits relative only to each other. Write an argument for or against the proposition that this latter assertion differs from the presentation in this chapter.

3.4. One often hears the statement, "that may be all right in theory, but it won't work in practice." Write an argument either for or against this proposition.

Hint: Would the meaning be changed if the statement were, "That may be all right in hypothesis, but it won't work in practice."

3.5. For the man-computer version of the example game given in Chapter II, explain the parts of the simulation that reflects each of the four properties of simulation models given on pages 27-28.

3.6. Think of an object system from your personal experience. Consider how you would create a model of it. State for this model what would be parameters, starting conditions, input variables, generated variables, time paths, and outputs. Briefly describe some of the operations of the model. Would you have to make this description of operations more specific for your simulation to run?

3.7. Draw a flow diagram similar to Figure 3.3 for your simulation model from Exercise 3.6. (Templates that make flow diagramming easy are usually available at book and stationery stores.)

3.8. Consider performing some experiments with your simulation model from Exercise 3.6. State how you would execute three different experiments with this model. What might you learn from each?

3.9. Assume your simulation model from Exercise 3.6 satisfies you as a constructed specific expression of a theory. State how you would execute an educational simulation using your simulation model. What might your live participants learn from their experience?

4

MAN-MODEL SIMULATION

Man-model simulation was illustrated in Chapter 2 by the game of matching pennies. The man component in that example was two live players. (Of course, the players might be female; the word *man* is used here in the sense of mankind.) The model component was the adapted rules of the game. This model represented only a part of the object system we said we were interested in. The man component made up the rest.

The distinguishing characteristic of man-model simulation is that some human counterparts of the object system are represented by live participants while any other human entities and the relevant nonhuman features of the object system are represented by a model.

Sometimes the model is conceived around convenient physical objects, devices, or locations. While in our example game the model itself (i.e., the adapted rules) is purely conceptual and the game can be played solely by communication between players, execution of the model is aided by the physical possession, exposure, and transfer of tokens, in this case the 100 pennies given each player at the start of each simulation run. When the device that aids execution of the model is a digital computer, we shall use the term *man-computer simulation.* Man-computer simulation is the subject of Chapter 6.

Many statements made about man-model simulation also apply to man-computer simulation. Depending on purposes or subject areas, man-model simulation and man-computer simulation are variously called operational gaming, business simulation, management games, educational gaming, bargaining games, experimental economics, political gaming, and the like.

One could think of the entire man-model simulation itself as a super model; in this view, a live human participant occupies a node for an operation and the "operation" at this point is obtaining a response from him. However, in this book we will restrict the idea of a model only to the constructed specific expression of a theory or of one or more hypotheses. We would not need to use live subjects in research if we had a good theory about their object system counterparts; instead we could construct a model of them and use the model. Hence, we need to use live subjects only for those counterparts we do not understand well enough to model. This use of live subjects is appropriate both when the object of research is human behavior itself in known systems and when it is the behavior of hypothetical systems given a kind of subject behavior that cannot presently be modeled. Live subjects in simulation studies are expensive; it is much cheaper to represent live persons by models.

In educational man-model simulation, we would not want to think that beginning students represent very well their experienced counterparts in the object system so we do not want to consider the students as an expression of our theory of their counterpart portion of reality.

Hence, for both research and educational reasons we do not think of live participants as operations in our models. A model, however, may specify nodes where behavior by live entities is an input that is prerequisite to the next operation prescribed by the model. A *live entity* may be a single person or group of persons responding as a single unit. Inputs at these nodes will function the same way in the model regardless of the number of persons making up the live entity.

Whole Simulation Systems

To formalize this distinction, we shall talk of the *whole simulation system* and various parts of it. A simulation model is one of the parts of a whole simulation system; it is specifically that part of the system intended to represent all or part of an object system. The rest of the whole simulation system consists of any live entities representing counterparts in the object system and the plans and activities of the researcher or educator and his staff for carrying out the simulation runs. Hence we create a system—the whole simulation system—to study or learn about another system—the object system—which we may or may not have had a hand in creating.

It will be convenient to speak of *segments* of a whole simulation system. These segments may contain part model and part human behavior. We should always remember that it is the whole simulation system that is used by decision makers and educators in science, industry, and government to pursue objectives external to the simulation and that the results of a simulation study must be interpreted in the context of the whole simulation system.

A *simulation* is the execution of a whole simulation system, which simply means that simulation as defined in Chapter 1 does not run by itself.

Flow Diagrams for Man-model Simulation

In flow diagrams, the behavior of a live entity is represented by a rhombus. A flow diagram of the man-model simulation of our example game from Chapter 2 is shown in Figure 4.1.

Notice in Figure 4.1 that the flow of operations goes in two directions simultaneously after the trial number counter is incremented. Ordinarily in flow diagrams this apparent ambiguity is not permitted. In our example game, however, we have more than one person or facility for providing behavior responses so that these responses may occur together in time. In live coin-matching games, for example, it is customary that exposures are made simultaneously

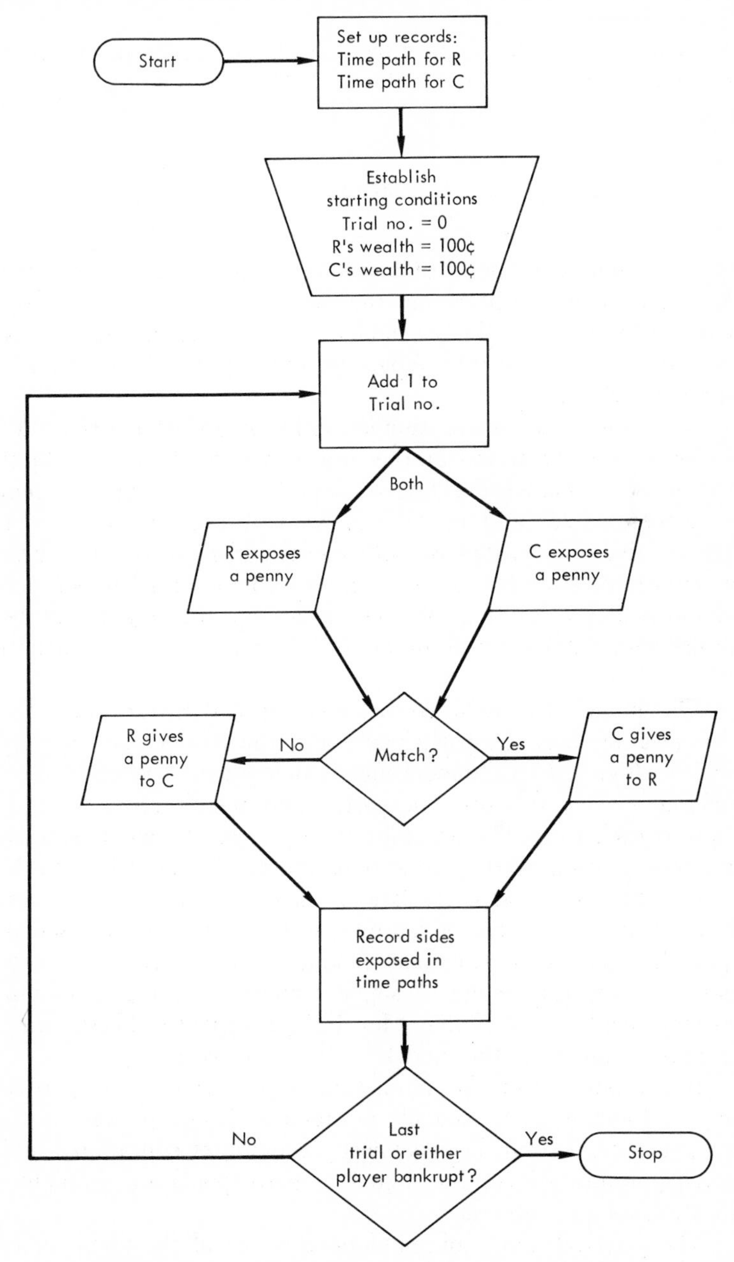

FIGURE 4.1. Man-Model Simulation of Example Game

in order to preserve the security of each player's decision. We note this simultaneity by writing "both" or "all" at the branching point.

Design of Man-Model Simulations

The essential property of man-model simulation is the interaction between live entities and the model. The whole simulation system is comprised of (1) the model, (2) live participants who assume the roles of some counterparts in the object system, and (3) administrative activities.

Designing a man-model simulation involves choosing the goals and objectives to be served by the simulation, specifying the object system to be simulated, deciding what portion of the object system will be represented by a model and what portion by live participants, specifying operations by both model and man to make these representations, establishing the forms of communication between man and model, describing the type of participants desired, and developing administrative procedures for carrying out the simulation runs.

The model, the required behavior of participants, and the communication between participants and the model may be very simple. In Figure 4.1 the model merely determines wins and losses, payoffs, and when to stop. The participants merely respond heads or tails at each turn. The administrative part of the whole simulation system provides starting conditions and keeps time path records.

In other man-model simulations such as management games, the model may be extremely complex, several pages of reports may be provided, and participants may conduct formal meetings of team members before responding. A single response in complex simulations may be a long list of values for decision variables. These values then become input for the model for the next cycle.

If the purpose of the simulation is educational, the participants are thought of as students or trainees. If the purpose is research, they are looked on as subjects. Sometimes educational and research purposes are combined so that participants are simultaneously students and subjects.

The practical design of simulations is one of the main topics of this book. We have only begun to touch on the considerations involved. In the next sections and throughout the remaining chapters, many aspects of simulation design will be covered.

It is appropriate at this point, just as we begin the study of simulation design, to emphasize that the design phase is critical to the success of a simulation. Frequently, simulation practitioners, after considerable resources have been expended, wish they had spent more of their time in the design phase before dashing into actual simulation runs. Where modifications are possible during or between simulation runs, these adaptations are essentially a return to the design phase.

The design and execution of a simulation project involves three distinct phases: 1) the design phase; 2) the execution or running phase; and 3) the evaluation or report phase. The person who oversees the second and third phases is called the administrator; he may or may not be the simulation designer from phase one.

Stimulus-Response Framework

It is convenient to think of the communication in man-model simulation as stimuli and responses. This orientation is frequently abbreviated *S-R*, meaning stimulus-response. We usually think of the live participants as the component that is stimulated and that responds. However, in all-computer simulation, we may want to think of the model, or model segments, as receiving stimuli and yielding responses.

Psychologists in conducting decision experiments with live subjects usually think in terms of a three-step communication sequence. First they present a stimulus, say by asking a question ("Will your opponent choose heads or tails on this trial?"), second they obtain a response ("I believe he will choose tails."), and third they announce the consequences of the responses ("He chose heads. You lose one penny.") This cycle is abbreviated S-R-C. When strung together for repeated cycles, we have S-R-C-S-R-C. . . . Since the experimenter performs both C and S, and since these are next to each other, C and S may be combined into a single step. This step is what the experimenter, administrator, or computer does between responses by the live subject. This single step is a combination of both current consequences and new stimuli, which we shall still call the stimulus (S) to retain our two-step stimulus-response (S-R) notion of communication between model and participant.

Participants receive stimuli from the administrator or experimenter as output from execution of the model. The stimuli are presented and the responses are returned to the model in a sequence that is fully stated in the flow diagram of the whole simulation system. The flow diagram should also show the points in this sequence where data are to be recorded by the administrator.

Time

Since we are excluding continuous-time, man-machine simulations such as flight trainers from this book, we restrict ourselves to time measured in discrete pieces, such as trials, cycles through the model, simulated time periods, or intervals of real time. The total simulated time represented by a simulation run is determined by the number of these discrete representations of real time that the model is operated. In our example game, total simulated time (i.e., number of trials) was a generated variable because we provided a stopping rule that depends upon the course of play of the game.

Man-model simulations are accomplished in sessions. A simulation run may last only one session, or, with multiple sessions, as long as several months. The number of sessions depends on the purposes and design of the simulation, the physical availability of the participants, the speed of processing the model, and the available resources. *Real run time* is the total length of all sessions in a run. Real run time does not include time between sessions. At first, participants require a long response interval. However, learning is usually rapid and response time soon approaches a constant level that can be used to estimate future real run times.

Pilot Runs

A pilot run of a man-model simulation is almost always needed to estimate participant response time. A pilot run also helps estimate model processing time. Pilot runs may uncover unforseen diffi-

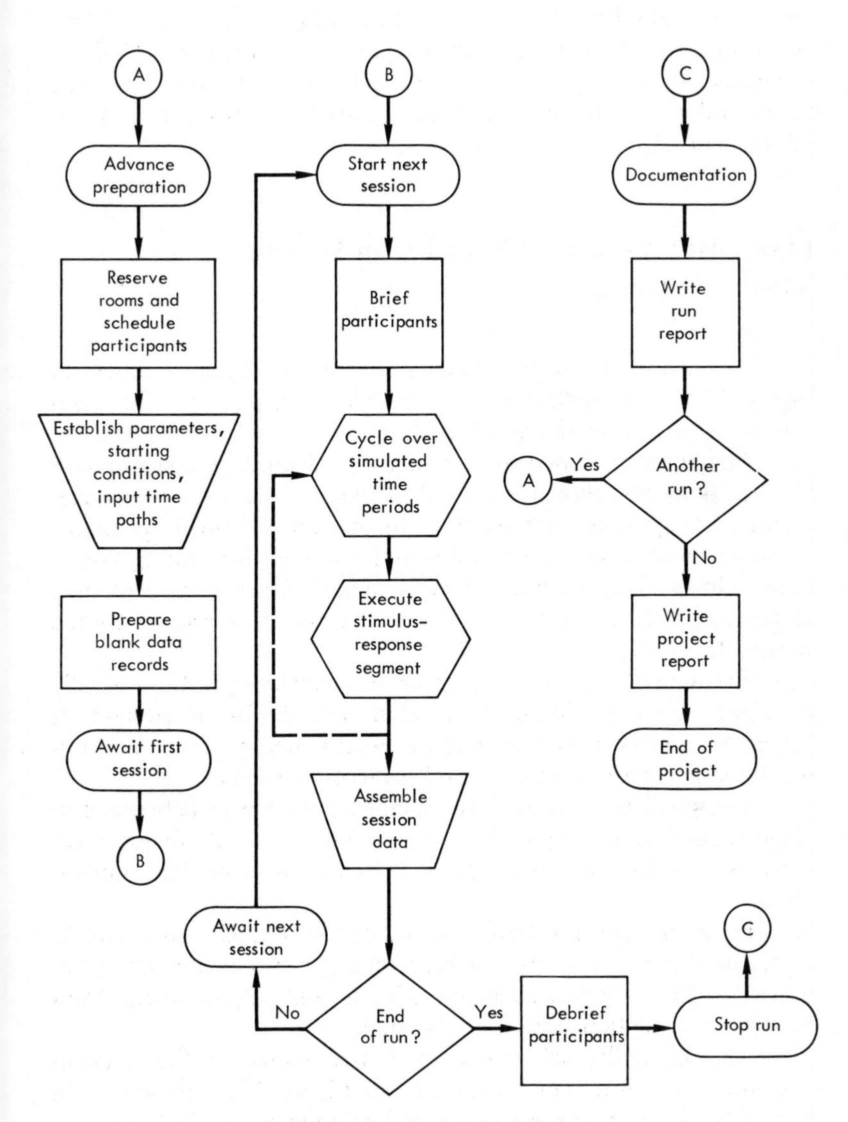

FIGURE 4.2. Man-Model Simulation

culties that could be expensive or embarrassing; examples of these are inconsistencies in processing procedures, unanticipated kinds of responses by participants, and odd or extreme results from the model. Pilot runs are frequently only partial runs using convenient rather than selected participants.

Flow Diagrams of the General Man-Model Simulation Design

A general flow diagram for man-model simulation is shown in Figure 4.2. New symbols introduced in this diagram are the hexagon, the square, and the small circle.

The hexagon is used to represent a predefined process or group of operations not shown in detail in that diagram. For example, cycling through time units requires the previously defined operations of incrementing the time counter and checking for ending conditions. This cycling is often called "looping." Operations contained in the loops are bracketed by a broken arrow that returns to the start of the cycle.

The square is used to represent an auxiliary operation, usually performed by the administrator. Examples shown by squares in Figure 4.2 are the initial briefing of participants by the administrator, his study of the final data, and his writing a report.

The small circle is used as a connector within or between flow diagrams; it does not represent an operation. A pair of labeled small circles shows flow just as an arrow does and is often less cumbersome.

Figure 4.2 also illustrates the use of the oval as a label and to represent starting, stopping, or interrupting the sequence of operations. As before, the trapezoid shows inputs and outputs to and from the whole simulation system or the model.

The undetailed set of operations represented by the hexagon "execute stimulus-response segment" in Figure 4.2 is presented in partial detail in Figures 4.3 and 4.4. These figures illustrate the use of ovals to tie simulation system segments together. In this use, ovals show the connection between a detailed segment and the hexagon for that segment in another diagram. The first oval of a segment serves as the label referred to by the hexagon. The word "return" in

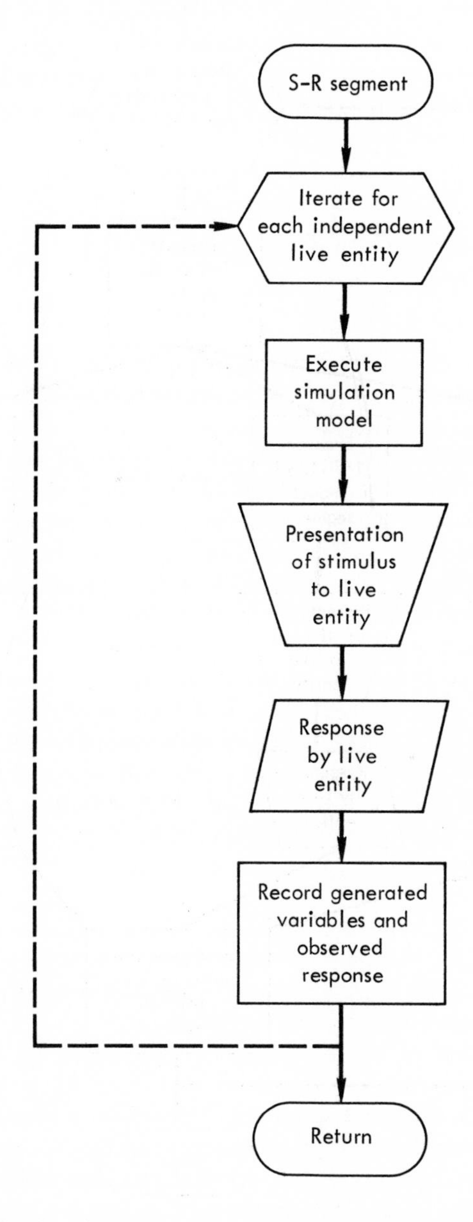

FIGURE 4.3. General Noninteraction Stimulus-Response Segment

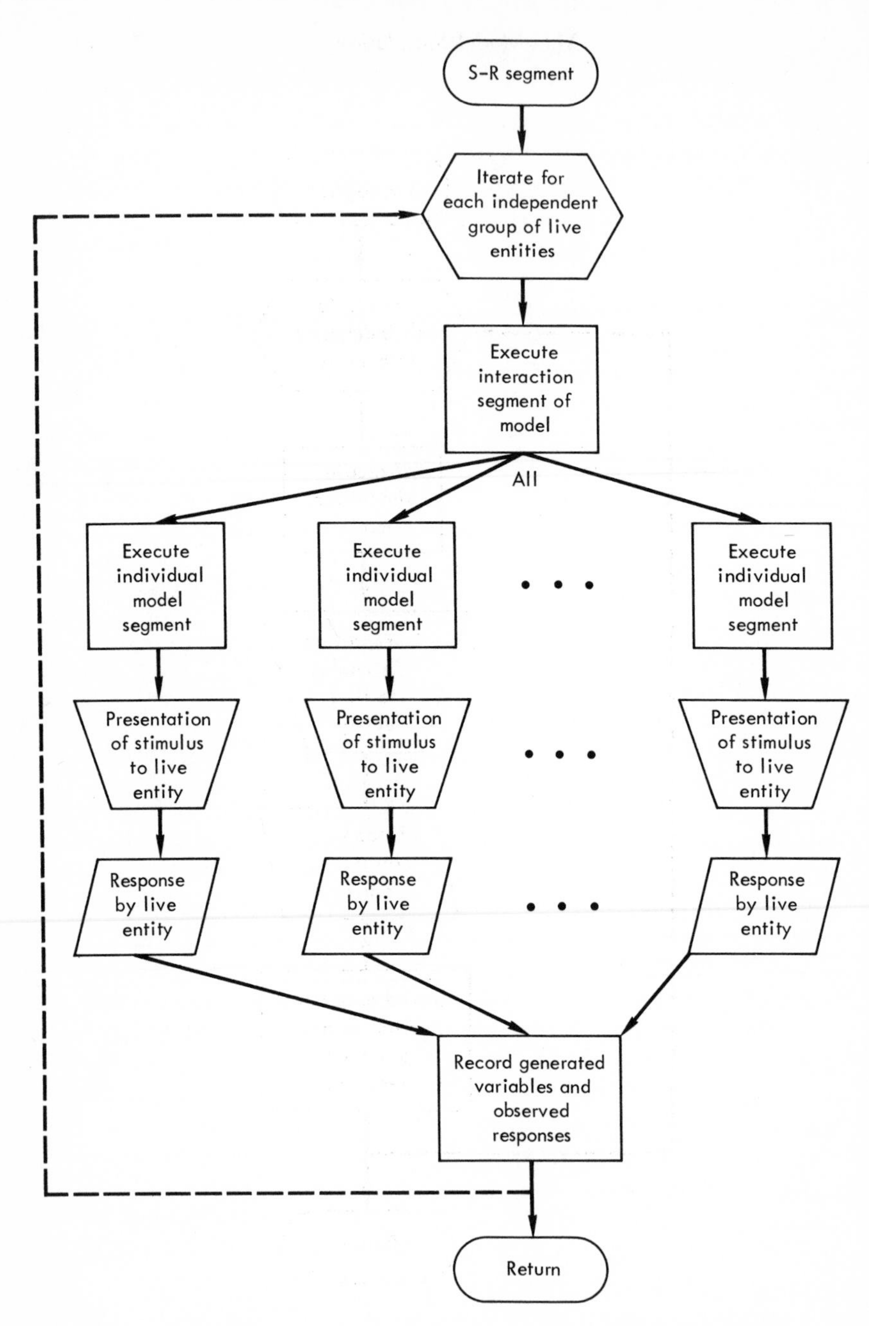

FIGURE 4.4. General Interaction Stimulus-Response Segment

the final oval means return to the main or referencing diagram and execute the next operation following the hexagon.

Interacting and Noninteracting Designs

In man-model simulation the stimuli generated by the model may depend on the behavior (responses) of more than one live entity; in this case, the simulation is called *interacting*. Interacting designs are also called *competitive* if more goal achievement by one live entity means less goal achievement for another and *coordinative* if two live entities can simultaneously obtain more goal achievement: Some designs contain aspects of both competition (conflict) and coordination (cooperation). If the stimulus to each live entity is independent of the responses of all other live entities, the simulation is called *noninteracting* or *noncompetitive*.

Figure 4.3 shows a general noninteraction stimulus-response segment. In this segment the same model is used for all live entities so that if this segment were referenced by Figure 4.2, the model would be executed once for each separate live entity for each simulated time period.

The repeated application of a model or segment is called *iteration*. Iteration, cycling, and looping are essentially the same; all are shown by a broken arrow that returns to the starting operation. We shall distingush what we are cycling or iterating over by using *cycling* with respect to time periods and *iteration* with respect to live or simulated entities. When Figures 4.2 and 4.3 are combined there is a nesting of an iteration within a cycling, i.e., a loop within a loop.

The simplest man-model simulation design is a noninteraction segment run for one participant for one simulated time period. This would be represented by one pass through the flow diagrams of Figures 4.2 and 4.3 with no looping back on the broken lines.

Figure 4.4 shows a general interaction stimulus-response segment. Any number of live entities may interact through the model, shown by three dots (an ellipsis) used to mean something is left out. If only one live entity is considered, there is no interaction through the model and Figure 4.4 reduces to Figure 4.3.

Interacting live entities form one interacting group of such entities. Parallel but independent groups may be run concurrently in the same sessions. The advantage of this procedure is that it enables processing of additional runs of the simulation without requiring additional sessions. Of course, more space and more administrative assistance may be needed. Figure 4.4 when embedded in Figure 4.2 requires that simulated time be the same for all independent groups. If iterating for different simulated times for each independent group, which means each may work at its own pace, simultaneous but independent sessions may be organized. This implies the groups may finish at different real times.

The individual stimulus model segment executed for each live entity within a group need not be identical. Different segments may produce stimuli of different kinds for specified live entities. This would be appropriate when each live entity represents a different role in a simulated organization. For example, in a coordinative simulation the entities may represent the several officers of a corporation where each officer receives reports unique to his role. This is in contrast to live entities that represent entire business firms in competitive interaction, where the same model segment is used to produce reports for each firm after the market interaction model segment is executed.

So far, the design of man-model simulation has been discussed in general terms. A specific design that can actually run requires the organization of five kinds of ingredients: 1) the participants, 2) the model, 3) the administrator's role, 4) the stimuli, and 5) the responses. We shall discuss the first three ingredients in this chapter and the last two in Chapter 6.

Participants

Ideally, the researcher would like to have subjects who are the actual counterparts in the object system, and the educator would like to have students who by reason of prior training will soon move into the counterpart roles in the object system. Such optimal participants are seldom available, usually because the administrator does not have complete control over selection.

Most man-model and man-computer simulations must be adapted to the participants that appear. One way this is done is by organizing participants so that special skills or backgrounds are distributed as desired. Other methods are making changes in the model, adjusting the behavior of the administrator, and modifying the stimuli and required responses.

SELECTION

However, administrators usually have partial control over participant selection. When students instead of real-life counterparts are used in research simulations, the administrator may enjoy a wide selection of student characteristics if he pays for student time and a narrower range of choice if he cannot pay but requires their services in connection with a classroom project. In educational simulation, curricular considerations usually determine which students experience particular simulations.

Educational simulations may also be run as controlled experiments. This may provide a type of research subject not otherwise available. For example, businessmen frequently play management games as part of executive development programs but are unwilling to serve as subjects in research projects due to time constraints. When games played by experienced managers are also conceived as research simulation runs, actual businessmen become research subjects.

With sufficient selection of participants, the administrator may be able to isolate a control group. The members of this group might receive a standard run of the simulation while another group, the experimental group, receives a different run. Or the control group may be selected so that it possesses a standard background or level of skill while the experimental group possesses some other characteristic, with both groups exposed to the same simulation run. In either case, it is desirable to match the groups as closely as possible in characteristics that are not being varied so that differences between the groups can be attributed as much as possible to the change in the simulation or to the change in the controlled difference between the groups.

Some participants may fail to appear at later sessions. Spare participants, who are redundant yet are completely participating during the run, enable the administrator to make substitutions as needed without injecting naive persons into the simulation run.

STOOGES

A special kind of live participant is the stooge. A stooge is in all appearances an ordinary participant, except that his behavior is controlled by the administrator in ways unknown to other participants. For example, in a simulation study of implementation of optimal policies, a stooge may offer an administrator-supplied optimal solution to the simulation problem; of interest is whether other participants recognize the stooge's suggestion as optimal and, if they do, whether they implement it. The behavior of stooges in research simulations is an experimental variable and in educational simulations a feature of the learning experience.

PRE-RUN INSTRUCTIONS

The amount of pre-run instruction given to participants varies with the purpose of the simulation. In some cases, participants may receive extensive specific training for the roles they are to play; in other cases they may learn the mechanics of the simulation only as the run progresses. *It is extremely important in both educational and research simulation that participants receive enough early instruction so the simple mechanics of their roles do not interfere with the behavior to be observed or the behavior that is the learning experience.*

ORGANIZATION OF LIVE ENTITIES

Participants as live entities may be organized in several ways. The model can operate whenever the required responses are received regardless of what constitutes a live entity.

Single persons may be live entities in three ways: 1) they may interact with the model independently, 2) they may interact with the model and with each other competitively through the model, e.g., our example game of matching pennies, or 3) they may interact through the model coordinatively with others to form a team. In the last case, the members of the team communicate with each other and with other teams only through the administrator.

Live entities may also be face-to-face teams. In this case, responses must be agreed upon by the team before being submitted to the operations of the model. Here the participants interact with all the nuances and conflicts of face-to-face communication. Face-to-face teams, each as a live entity, may be organized to interact with the model independently, or with the model and with other teams through the model. Many combinations are possible and the design chosen depends on the research or educational objectives of the simulation.

PHYSICAL CONTROLS

How the participants are organized determines the physical controls required during and between sessions. Single persons may be isolated in separate booths or rooms. Face-to-face teams may be similarly isolated. Isolation inhibits direct communication among participants and permits presentation of different briefings or stimuli.

When a run requires several sessions, uncontrolled communication among participants between sessions may invalidate the research or inject undesired collusion, as we suspected for Pair 3 in the man-model simulation of our matching pennies example in Chapter 1. To reduce unwanted communication the administrator may keep secret the real-person identity of live entities. This may require separate briefing rooms, separate briefing times, or separation of participants prior to sessions. Man-model simulation sometimes requires a large amount of floor space and numerous partitions. Special laboratories that contain briefing rooms, session rooms for individuals and for teams, and a variety of observation and communication equipment, have been constructed to accommodate man-model and man-computer simulations. These are usually called behavioral science, decision research, or management simulation laboratories, and they may also be used for other purposes.

The Model

The model is the main source of stimuli in man-model simulation. It represents that part of the object system that is not represented by live participants.

VERISIMILITUDE

Since the model represents to participants aspects of object system reality, it must give to them the appearance of or the effect of reality. Simulation does not require reproduction in every detail, but it does require capturing the relevant aspects as the participants see them. This effect is called *verisimilitude,* which means that, while the simulation is obviously an artificial representation, it has the quality of being true to life or to human experience.

The model must be complex enough to provide verisimilitude, yet be simple enough to process in the available time. Computers by speeding up processing enable designers to build more complex models.

FLEXIBILITY

The model should be flexible enough to accommodate parallel runs. This is especially important when control groups and experimental groups are run concurrently. Parallel or sequential processing of the model may execute runs that differ in parameters, starting conditions, input data, or even in parts of the model. Figure 4.3 is an example of parallel processing. The iterations over each independent live entity during each time period in Figure 4.3 represent as many independent parallel runs of the simulation as there are live entities. These concurrent runs may use identical input and model operations or they may use different input or model parts.

The model should also be flexible enough to adapt, either before or during a run, to the characteristics or behavior of the received participants.

INTERACTIONS

For interacting man-model simulations, the model will have one or more interaction segments. Hierarchies of interactions are possible. Groups of interacting live entities may interact first among themselves and then between groups. Once the interactions are processed for the highest hierarchal level, the results must usually be further processed for each group and then each live entity in a group before being fed back as the next stimuli.

SEQUENTIAL DEPENDENCE

If stimuli or responses from one cycle influence the stimuli in subsequent cycles or periods, the design is said to include *sequential dependence*. In the penny matching example, knowledge by a live player of the past behavior of his opponent provides sequential dependence if this information influences his choices. Most man-model simulations are sequentially dependent in order to reflect the dynamic aspects of the object system. The input at each cycle for a sequentially dependent model may be some or all of current or past live responses or generated variables. Sometimes the entire history to date of a run is reflected to some degree in each cycle.

DOCUMENTATION

Every model should be documented in such a way that another researcher or educator can use it. Documentation may include a statement of the step-by-step procedure for processing the model, a detailed description of the model, a flow diagram, data recording details, and a description of the forms of stimuli and responses.

The Administrator's Role

In research simulation the administrator is usually one of the designers of the model, but in educational simulation, as each simulation becomes widely used, the personal attention of the original designer moves farther and farther from actual runs. Thus a clear documentation of the administrator's role is more important in educational simulation than in research applications. On the other hand, bias in results is more sensitive to the administrator's behavior in research runs than in educational runs. The following duties of administrators apply to both research and educational man-model and man-computer simulations. The importance of these duties varies with the purpose of each simulation project.

OBTAINING PARTICIPANTS

First, of course, participants must be obtained. Participants may appear as part of another organized activity such as enrollment in a course, attendance at a conference, or participation in an executive development program. Or they may be recruited through employment agencies, through visitations to classrooms, by announcements, or through classified advertisements. Care must be taken to clearly state during initial contacts with participants the obligations to be imposed on them. Misunderstood recruiting promises can affect behavior during a run.

PRELIMINARY ARRANGEMENTS

Scheduling of the run may be done before or after participants are contacted, depending on the source of participants. Arrangements for rooms and other physical facilities must be completed far enough in advance to notify participants when and where to appear. Transportation and meals may be provided. Arrangements for proctors and assistants need to be made.

Advance written instruction, if any, can be given to participants when they are recruited, when they are scheduled, or mailed to them prior to the first session. Directions about the amount of advance reading and study should be specific. Once participants are known, adapting the run to their special characteristics should be considered.

RECEIVING PARTICIPANTS

At the first session, participants must be received and guided to their rooms, booths, or tables. They should be reassured at this time about their commitments and their earnings (if any). Most designs provide for a briefing session prior to the simulation run. Sometimes the briefing is continued into the first simulation session and combined with practice trials. If questionnaires or tests are to be completed prior to a run, this is most easily done in conjunction with the initial briefing session.

RUN SUPERVISION

Supervision of participant behavior during the actual run may require proctors. Assistants to process the model may be needed. These must be recruited and trained in advance. During a run, the administrator should check that the proctors and model processors are performing as planned.

The amount of instruction and advice given by the administrator and proctors during a run should be carefully planned and controlled. Too little assistance to participants may prevent some from learning the mechanical details thus impeding their experience with the substantive content of the simulation. Too much advice to participants may influence their behavior in unwanted or even unobserved ways. A system for signaling for assistance by participants should be established and explained. If there is more than one session in a simulation run, at the beginning of each, rebriefing may be conducted and substitutes for missing participants arranged.

Sessions in some applications may be very informal. For example, in one educational technique, the teacher acts as administrator for only a prescribed number of minutes of class time, usually after reports or other stimuli are delivered. He then sets a deadline for turning in written responses at some separate location. During intervening classroom contacts he may not again discuss the simulation until the next stimuli are delivered.

When participants are geographically dispersed and communicate with the administrator by mail or over telecommunication lines, a session for a participant begins when he receives his stimulus and ends when he sends back his response. In this situation, the stimuli must be self-administering.

The administrator should be prepared to meet emergencies of many kinds. Only highly refined man-model simulations such as often-run educational simulations are relatively free of unexpected consequences. A pilot run will provide experience in handling unexpected circumstances, but each batch of participants will also provide new situations. Untoward results in research simulations may mean that some or all data must be discarded. In educational simulation, the administrator should secure frequent feedback to judge whether participants are making the desired inferences; if not, he should attempt to lead them in their discovery processes; as a final resort he may simply state the principles underlying the simulation exercise.

ENDING PROCEDURE

Debriefing after the last session can serve several objectives. Requests not to discuss certain features with outsiders can be made or repeated. Written or oral questioning of participants may provide administrative insights into response processes. Final data may be gathered and missing items can be checked. Extensive individual interviews may be conducted. In educational simulations, discussion about relative performances provides a final opportunity to lead participants to desired inferences.

All commitments to participants must be fulfilled. Promised payments should be made immediately after final debriefing. Arrangements should be made to mail reports of results, if promised.

Following debriefing, the administrator should immediately prepare a formal "run" report for his records. This report should be designed when the simulation is planned. It should also include unusual circumstances encountered and suggestions for improving future runs.

Finally, following completion of runs, for research projects the collected data may be analyzed and a report written; for educational applications, a review and evaluation may be prepared.

EXERCISES

4.1. Prepare a flow diagram for a man-model simulation of the example penny matching game using the sophistication of Figures 4.2 and 4.4, but only as much as appropriate.

4.2. Prepare a stimulus-response segment flow diagram for a three-person man-model simulation of a bidding object system in which each bidder is one of three automobile repair firms. These firms bid for repair jobs that vary in size and are to be paid for by insurance companies.

4.3. Design a complete man-model simulation for the automobile repair bidding object system of Exercise 4.2.

4.4. Conduct a pilot run of your design for Exercise 4.3. Estimate real run time.

4.5. Execute a complete run of your simulation design for Exercise 4.3.

4.6. Design a noninteracting man-model simulation, including a complete flow diagram, to study the behavior of individual bidders in the automobile repair object system of Exercise 4.2. Plan to study each bidder separately over a sequence of bids. Let the model represent his competitors.

4.7. Execute a complete run of your simulation design for Exercise 4.6.

4.8. Design an interacting simulation of the automobile repair object system of Exercise 4.2. Let each bidder be a team, but require that team members contribute to team decisions only through the administrator by means of paper and pencil.

4.9. As a class or group exercise, execute a complete run of a design for Exercise 4.8.

4.10. Write a report on a design you have written for one of the above Exercises 4.3, 4.6, or 4.8. This report should show how the simulation may be used separately for education, separately for research, and concurrently for both education and research.

5

COMPUTER SYSTEMS

Man-computer simulation, which we shall discuss in the next chapter, differs from man-model simulation by the use of computers to execute models. This change provides a host of new opportunities for research and education. It also adds some constraints.

One major difference computers provide from man-model simulation is the almost unbelievable speed in processing the model. In some applications, stimuli following live responses are fed back to participants so fast they may hardly be ready for them. In others, delays in obtaining the services of a computer may slow stimuli for the next cycle so much that participants are distracted from their simulation roles, despite the lightning speed of processing once the simulation model is loaded into computer memory.

Another difference computers provide is in the physical characteristics of the stimuli and the manner of receipt of responses. Many variations are possible depending on the size and speed of the computer and on the kind of input-output devices available. For these reasons we have delayed until the next chapter discussion of the forms of stimuli and responses in man-model simulation so we can include man-computer considerations as well.

In this chapter we begin the study of the role of computers in both man-computer and all-computer simulation by briefly reviewing the characteristics of computer systems. In later chapters we shall consider problems of computer model construction, investigate the extraordinary capacity of computers for data collection and manipulation, and finally, see how the unique abilities of computers have literally extended the mind of man.

We shall restrict our attention to the use of stored-program, general-purpose digital electronic computers. This is the class of machines that is proliferating throughout government, industry, and education.

More on Whole Simulation Systems

In man-model simulation, the model represents that part of the object system not represented by live entities. The balance of the whole simulation system consists of the administrator, schedules, data forms, proctors, gathering of responses, delivering of stimuli, and keeping records. In man-computer simulation the model is written in one of the computer languages and is appropriately called a *computer simulation model.* The computer simulation model is only a part of the total computer program. The program can also cause the computer system to take over such duties as gathering responses, delivering stimuli, and keeping records; the computer can even aid in administering the simulation. All this is in addition to executing model segments that represent portions of object systems.

The structure of man-computer simulation can be illustrated by a successively inclusive hierarchy of subdivisions of the whole simulation system, as shown in Figure 5.1. This same hierarchy also applies to flow diagrams so that we may have flow diagrams of the simulation system, of the computer program, and of the model—all of which could be shown together in one super-sized flow diagram. Instead of one giant diagram, we draw flow diagrams in manageable

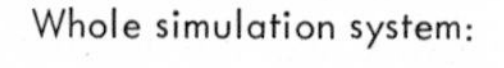

Whole simulation system:

1. Administrator, assistants, and their duties
2. Participants (if any) and their roles
3. Physical arrangements (including input-output devices)
4. Computer service including any computer-assisted administration conducted by the --

Computer simulation program:

1. Control segments
2. Data collection, reduction, and reporting segments
3. Stimulus-response segments, including communication with participants (if any) and the --

Computer simulation model:

1. Interaction segments (if any)
2. Individual stimulus generation segments

FIGURE 5.1. Subdivisions of Computer-Based Simulation Systems

sizes of the separate subdivisions and of convenient segments using circles, hexagons, and ovals to show the way they fit together. When live participants are omitted, Figure 5.1 applies as well to all-computer simulation.

Computer Languages

A simulation user of modern digital computers need not be a computer operator or an expert in programming computers in their own individual wired-in capabilities. Special languages, called *problem-oriented languages*, have been developed that look like familiar languages rather than like the actual operations of the machine itself. A *computer program* is a set of instructions that a computer first stores in its memory and then proceeds to execute. When under control of an internally stored program, a computer literally instructs itself what operations to perform next without any further help from the computer operator.

When a user writes a program in a problem-oriented computer language, he must follow exactly the rules of that language. If no mistakes have been made, the computer itself can translate the user's statements into equivalent statements in its own machine language. This process is called *compiling*. If compilation is unsuccessful due to mistakes by the user-programmer, the computer sends him messages that pinpoint his mistakes so he can correct them. The set of statements written in the problem-oriented language is called a *source program*, and the set of machine language instructions which the computer compiles and actually stores and executes is called an *object program*. A computer model written in a source language may run on many different computers despite the fact that the machine instructions the computer compiles from the source statements differ and are incompatible from one computer type to another.

Several source languages have been developed that are applicable to simulation problems. Some of these source languages are specifically created to compile simulation programs. One of these special simulation languages requires only that the programmer learn to draw a special kind of flow chart. These languages will be reviewed in Chapter 9. Until then we shall restrict ourselves to the language of flow diagrams. Flow diagrams are common to most source languages so that actual computer programs may be written from our flow diagrams. Thus we do not need to undertake learning any particular source language to pursue our present purposes.

Computer System Defined

Well established source languages, their compiler programs, and other service programs are called computer *software*. Computing machines and their peripheral equipment are called computer *hardware*. Peripheral equipment includes input-output devices such as punched card readers, card punches, magnetic tape handlers, typewriters, teleprinters, keyboards, light pens, and visual display units. A *computer system* is an operating combination of software and hardware.

Variations in Computer Services

Computer services are supplied to users in several ways. In one mode, called *hands-on* or *open-shop*, each user gains exclusive control of the computer and operates it himself. In another mode, called *hands-off*, *closed-shop*, or *batch processing*, the user gives his prepared programs and data to regular staff persons who operate the computer; later his materials along with the computer output are returned to him.

When a user has hands-on computer service, he may correct his programming mistakes or change an existing program the moment he sees a need to do so. He can then immediately try the program again on the computer. Thus, the hands-on user interacts with the computer in ordinary real time. In a sense, he is an adaptive human adjunct to the peripheral equipment that is attached on the electrical lines, or "online" so to speak, with the computer. Technically, *online* means the piece of equipment is monitored by or is under control of the computer. *Offline* means the connection between the equipment and the computer is by means of a hand carry. We could characterize the open-shop, hands-on mode of computer service as *online real-time* processing, but these terms have come to mean something slightly different from a user's point of view and something dramatically different for designers of computer hardware and software.

To illustrate this meaning, imagine a user having immediate hands-on real-time service (as if he controlled the entire computer himself) but at a remote location where only one or a few pieces of peripheral equipment are actually handled by him. Remote locations so equipped are called *remote stations* or *inquiry stations*. A unit of online equipment at a remote station is called a *remote terminal* or *remote console*. When this one user exclusively controls the computer from his remote station, the situation is merely the equivalent of having long cables from the open-shop computer to some input-output equipment, except that perhaps a telephone call to the computer location may be required to have the computer turned on or a magnetic tape changed.

Now imagine many such users at different remote locations all receiving online real-time service simultaneously from a centrally located computer. Also imagine that the computer is so much faster than the users that it keeps them all as busy as they wish to be (or even busier). Thus the effect at each location is as if each user had his own computer. A computer system that provides such service is called a *time-sharing* system since many users simultaneously share the real time the computer is running. However, not all systems claimed to be time sharing are fast enough to avoid some waiting by users at terminals.

Historical Review

If we look back on the development of electronic computing services in a brief historical review, we see that users originally had exclusive access to a computer for a limited amount of real time; these first users wrote their programs in machine language. Later, the computer room or shop was closed and computer users were evicted. A user's program was then run by a staff operator in a batch containing programs of other users. If something was wrong when his output was received, the user needed to turn in his program again. (The activity on one computer job between turning it in and receiving it back came to be called a *turnaround.* The length of time required for this activity is called *turnaround time.)* Sometimes many turnarounds were required before all the "bugs" were out of a program. (A program that runs as desired is said to be *debugged.)*

Parallel with the development of closed-shop processing of programs in batches, almost as if to compensate users for long turnaround times and the absence of hands-on debugging privileges, source languages and compilers were developed. These made the tasks of programming computers easier and more reliable. Now the development of time sharing has almost returned hands-on rapid-turnaround service to users, but with the added advantages over old hands-on service of easy and reliable problem-oriented programming languages, faster and larger computers, and access to the computer from remote locations. All three modes—hands-on, batch-processing, and outline real-time service—are currently prevalent and used in both man-computer and all-computer simulation.

The most recent innovation, which is a blend of both software and hardware, is *multiprogramming*. Multiprogramming is a mode of central computer operation in which the computer contains more than one program at once and executes them either by finishing each in turn or by taking brief turns called *time slices* on each program. The ultimate in multiprogramming is *paging*, in which the computer organizes programs within itself into small segments called *pages* that are called into memory as needed. These developments mean that one central computer can communicate in parallel with many users or participants at terminals while almost simultaneously processing their programs or responses, and in addition, during moments of low demand from remote terminals, process programs turned in at the computer center.

The implications of online, real-time, time-sharing, multiprogramming capability for man-computer simulations are vast and have hardly been explored. The prospects for both research and education are exciting, but they also present new issues in the design of man-computer simulations.

Input-Output Equipment and Flow Diagram Conventions

When we include a computer in simulation systems, we need to describe the means of input to and output from the computer. Also, we need to make distinctions in our simulation designs between types of input-output devices. This is conveniently done with flow diagram symbols. Since we are not learning a specific computer source language in this book, flow diagrams are serving us as a language in which our simulation system designs, models, and computer programs are written.

All of our previous flow diagrams conventions will remain meaningful when the model is executed by a computer. The symbols for performing an operation (rectangle), making a decision (diamond), and executing a predefined process (hexagon) will now also represent operations performed by the computer. With computers, assistants may still be needed to prepare computer input, pick up and distribute computer output, proctor, and otherwise aid in administering the simulation run.

Input-output in flow diagrams has therefore been represented in general by a trapezoid. We shall now describe specific types of input-output equipment and introduce their flow diagram representations. We start with output devices since through them flow the stimuli that stimulate part of the object system.

Output Media

Output from a computer may take several forms. One of these is the printed document, symbolized as shown in (a) in Figure 5.2. The symbols in Figure 5.2 are used both for the form of the output (paper document, punched card, etc.) and for the peripheral equipment that creates it (online printer, card punch, etc.). Documents may be printed by high-speed, online printers (usually located in the computer room and using paper about 14 inches wide), typed by a typewriter located on the console of the computer, typed by online teleprinters or teletypewriters located anywhere (typing devices use paper about 8½ inches wide), or printed offline from cards or tape. Narrow paper and slow typing are common at remote locations while wide paper and fast printing are usually found in the computer room.

Computer output may also take the form of punched cards, symbolized as in (b) in Figure 5.2. High speed card punches usually exist only in the computer room. Card output punched at remote locations ordinarily occurs at relatively slow speeds. Cards can be listed on offline tabulating machines to produce printed output.

A widely used form of computer output is magnetic tape, symbolized as in (c) in Figure 5.2. Data is stored on magnetic tape by means of spots that are magnetized according to a code. Magnetic tape output is common where batch processing is done by fast expensive computers. Usually these computers have restricted online equipment and most printing and card punching is done by a less expensive auxiliary computer from an original magnetic tape output hand carried to it. The purpose of using a small computer as an input-output device for a large computer via magnetic tape is to

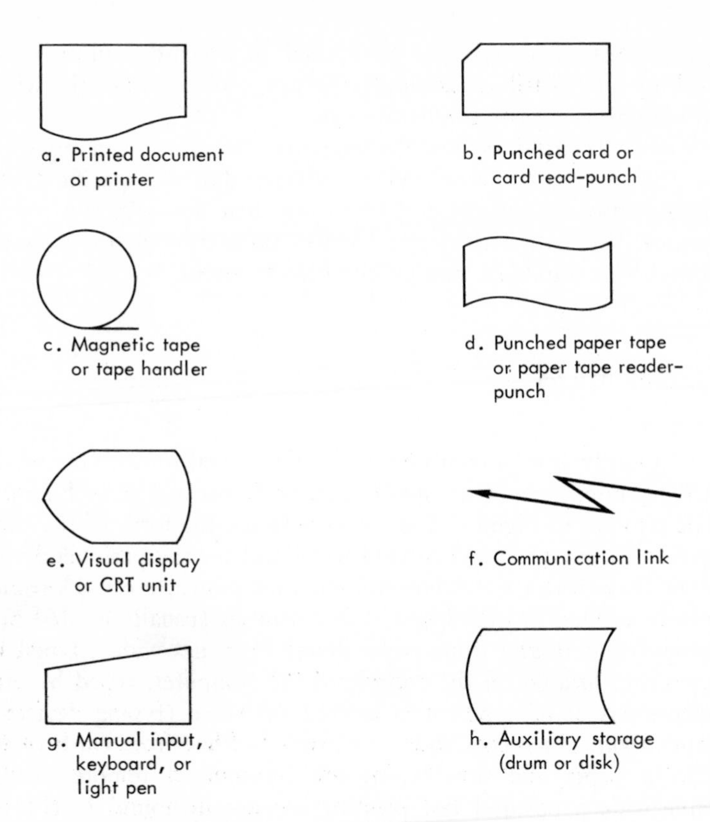

FIGURE 5.2. Input-Output Symbols

free the larger faster computer from card reading, printing, and punching chores that would slow it down; in this way each machine does only those tasks it performs most economically. Sometimes small computers are used as online terminals in time-sharing systems and the need to hand carry a tape is eliminated.

Punched paper tape is another form of computer output. It may appear in conjunction with a teleprinter that concurrently types the same data. With paper tape, the printed record can be reproduced by running the punched paper tape back through the teleprinter. The flow diagram symbol for punched paper tape is shown in (d) in Figure 5.2.

It is the holes punched in cards or paper tape and the magnetized spots on magnetic tape that contain the output; these forms and their codes are not easily read by live participants in simulations. The output they contain must be translated into meaningful displays before going to participants. Cards and tape may be used without this translation as temporary or permanent external storage for time paths and other generated variables.

Visual devices for computer output have recently been developed, notably cathode ray tube (CRT) display units that look much like television receivers. CRT displays are large or small, highly specialized or for general use, depending on their purposes and on the electronic equipment that generates signals for them. They are symbolized in flow diagrams as in (e) in Figure 5.2. Images generated on CRT displays may be mere duplication, both in appearance and speed of creation, of records that could also have been typed by teleprinters. On the other hand, a simulation user may wish to take advantage of the unique characteristics of CRT devices and by means of graphic techniques dynamically display illustrations such as maps and chessboards. CRT images, like those on a television screen, are temporary; in some systems what is called a *hard copy* (a permanent printed copy) can be obtained when requested if printing is also available in the computer system.

The symbol for a communication link between a remote station and the central computer is shown in (f) in Figure 5.2. Arrowheads indicate the direction of information flow; information flow may be one-way or two-way. Rates of transmission depend on the capacities of these links. Sending of digital information over ordinary telephone lines is slower than sending it through the large cables seen in computer rooms, but, for long distances, it is less expensive. In general, with respect to both transmission lines and input-output equipment, the faster the information flow, the more expensive the real-time dollar rate for using the facilities. Also, the greater the distance, the more the cost of transmission for a given quantity of information. This is in contrast to computer processing where the faster and more expensive per hour the computer, usually the less the cost for processing a given program.

Input Media

Of the above output media, punched cards, punched paper tape, and magnetic tape are also input media for computers. As input, cards and paper tape are initially punched on special offline devices that operate like typewriters. Characters on magnetic tape are written electronically by a card-to-tape converter or an input computer, or as carryover from a previous program on the main computer. Usually, a printed copy of what has been punched as holes in cards or paper tape or magnetized on a tape may be obtained.

Other input media are switches on the computer console, keyboards, and certain highly specialized devices such as arrays of buttons, light pens, movable writing arms, and metal tablets. Keyboards may be part of the computer's own typewriter or may be associated with a teleprinter or a CRT display unit at a remote station. A combination keyboard and teleprinter that will operate remotely is called a teletypewriter. Keyboards usually follow typewriter or adding machine conventions. Arrays of buttons or other switches have the advantage that descriptive overlays can show the meaning of each button or switch. Sometimes caps are placed over keys or buttons to change their identities. Keyboards and button devices may also be custom built for special applications. Light pens create input when an aiming beam of light is directed at predefined spots on a CRT display. Light pens offer an electronic analogue of multiple-choice paper and pencil responses. Movable arms and metal tablets as computer input for man-computer simulations are still in the experimental stage. All these forms of manual input to a computer are symbolized as in (g) in Figure 5.2. Usually this symbol will represent a keyboard or keyed-in responses.

The speed and cost characteristics described above for computer output also apply to input. The most common input in batch processing is punched cards and the most common form at remote stations is by keyboard. Keyboard input is temporary and is lost at the remote station when transmitted unless it is concurrently recorded on punched paper tape or on a hard copy. Card and tape input is returned to the user and can be used over and over again without further preparation. Various combinations of batch-processing input and output devices applicable to simulation are shown in Figure 5.3. Combinations available in online real-time systems are shown in Figure 5.4.

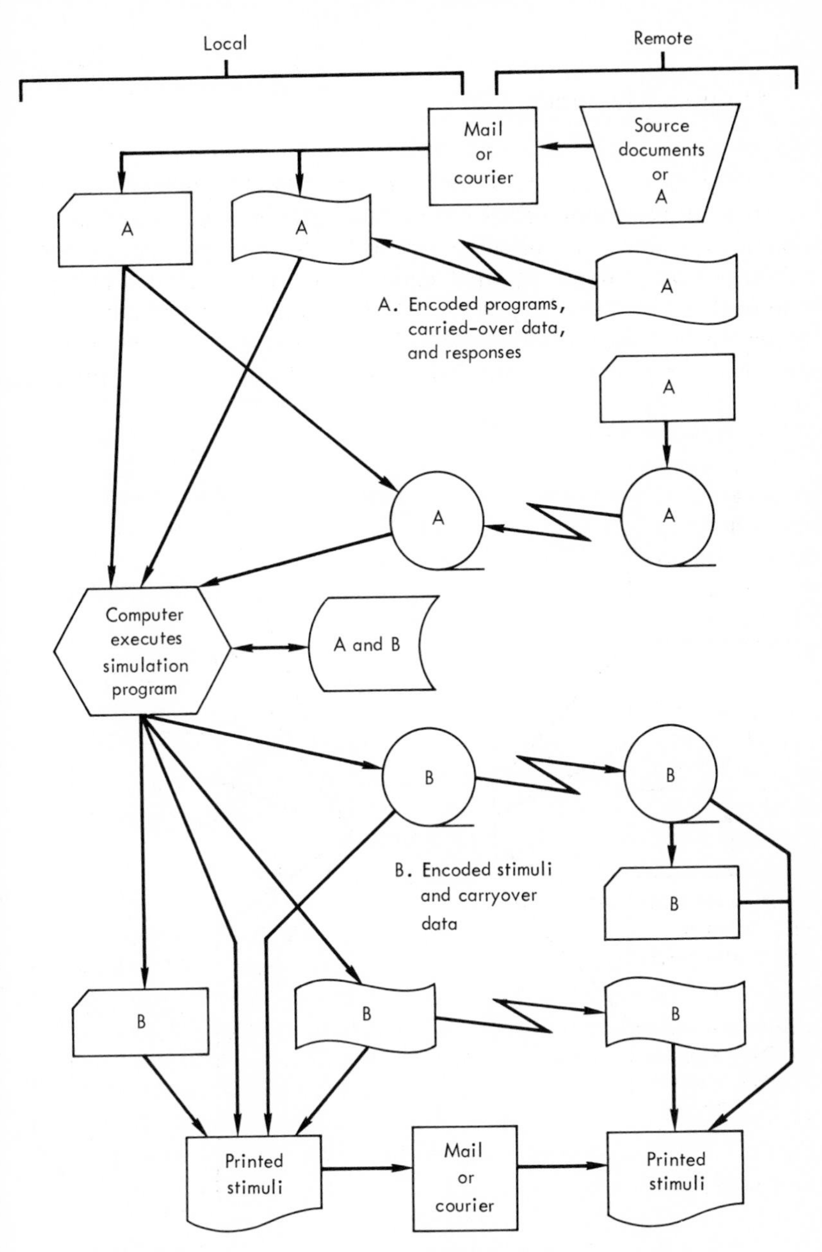

FIGURE 5.3. Closed-Shop, Batch-Processing Input-Output Variations

Computer Memories

Computers store programs and the data the programs use in their main memories, called *primary storage*. The control and arithmetic-logic units of computers operate only from primary storage. Storage that provides repeated rapid access (which cards and tape do not) is called *auxiliary storage*. Input, output, and programs may temporarily reside in auxiliary storage, but these must first be moved to main or primary storage before arithmetical or logical instruc-

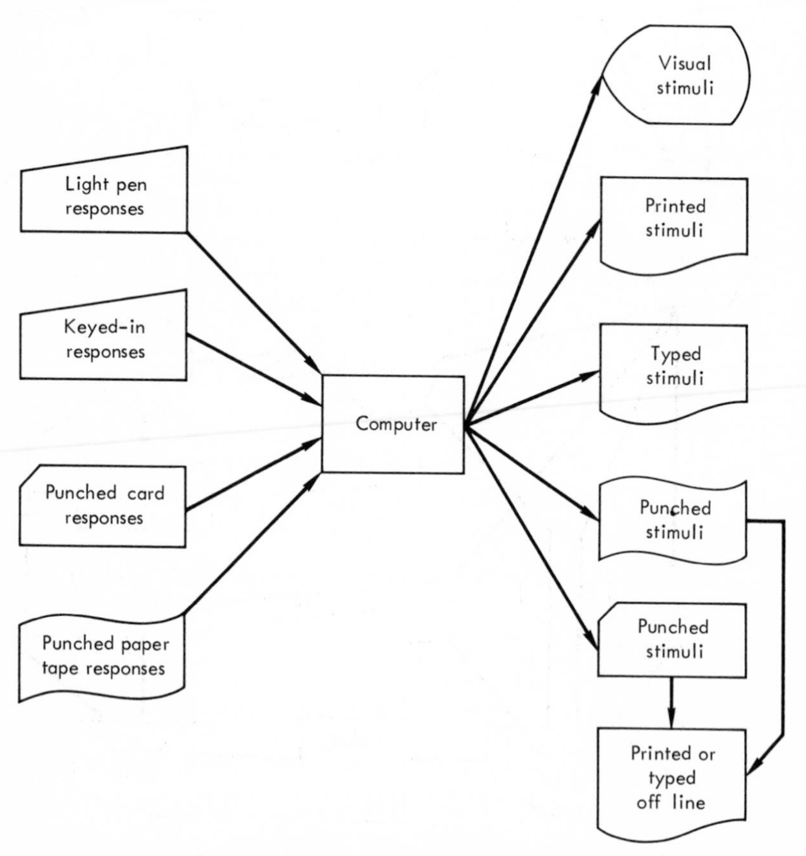

FIGURE 5.4. Open-Shop and Online, Real-Time Input-Output Variations

tions can be executed. Examples of auxiliary storage devices are magnetic disks and magnetic drums, symbolized as in (h) in Figure 5.2. Drums and some disks are permanently online; other disks are removable in "disk packs" that can be stored offline. These units can hold many programs and the data for each, any one of which may be rapidly loaded into the computer, processed, and unloaded again to the disk or drum, then moments later reloaded, processed again, and unloaded and so on. Other programs on the disk or drum may be loaded, processed, and unloaded while the first program is inactive. The rapid and repeated loading and unloading of different programs and their data bases, or parts of them, into and out of the computer is called *swapping*: swapping is widely used in time-sharing systems.

The computer itself controls its own work through special programs that moniter input devices, queue jobs, look for next jobs, find and report errors, interrupt jobs in process, perform swapping, assign and operate output devices, and so on. The set of these programs for a computer installation is called its *system programming* or *systems software*. Usually a portion of primary storage is permanently dedicated to a master systems program from which control of all other systems and user programs emanates. The computer even helps load its own master control program in a process called *bootstrapping*.

Computer User Behavior

At first computer users were their only operators. Later the following hierarchy of positions came into being:

1. User—person with a computer-related problem.
2. Systems analyst—person with methods of solving problems using computers.
3. Programmer—person who can code problem solutions to run effectively on computers.
4. Keypunch operator—person who transfers data and program codes to machine-readable form.
5. Operator—person who tends a running computer.

There are usually large differences in salaries among these levels, with the lower-salaried persons performing the keyboard communication function three levels away from the user.

Time-sharing is changing this scheme of user behavior. Of course, when an online remote terminal is a direct substitute for an offline keypunch machine, the hierarchy remains. But in time-sharing systems that interact online and in real time with persons at remote terminals, the persons sitting at the keyboards tends to become the more highly-paid systems analysts and ultimate users.

As scientists and managers more and more communicate directly with computers through time-sharing, a new hierarchy may appear:

1. Principal user—scientist or manager who understands the computer models he uses but may not be skilled in the keyboard procedures of the time-sharing system.
2. Time-sharing operator—person who is skilled in detailed time-sharing user procedures and to whom principal user can dictate instructions and data for keyboard input at the remote terminal.

Principal users may even dictate instructions over the telephone or into dictating equipment. Responses from the computer may be displayed or printed at the remote terminal, or they may appear on separate television screens observable by the principal user at *his* location, which may be some distance away from the terminal and time-sharing operator.

As time-sharing service broadens, time-sharing operators may become a regular part of the scene at executive meetings, in man-computer simulations, or in other settings where combination voice and visual communication with computers is desired.

At meetings of computer professionals, it is sometimes facetiously predicted that a future user will need only to sit at a terminal and utter grunts, "Uh-uh" or "Unh-unh." However, computer professionals are known both for promises that are kept beyond predictions and for some that are not.

Using a Computer System

Certain maxims have developed for the efficient and tranquil use of computer systems. These suggestions are most applicable to closed-shop computer service where you—the user—are entirely dependent upon staff operators to handle your programs and data as you wish. However, the maxims contain sound advice for hands-on and online real-time users as well. The maxims are listed below:

1. Assume the people who operate and maintain the computer system do so to serve you as a user. They want your jobs to run properly.
2. Expect a large number of errors from yourself.
3. Expect a few errors from the system. Report these immediately to the systems supervisor so he can correct them. There is a delicate line between reporting errors that might have been caused by you to the systems supervisor, only to have your own mistakes pointed out, and reporting errors that are legitimate systems difficulties that the supervisor urgently wants to know about. Systems supervisors depend on the reports of users to discover malfunctions, but they do not have time to defend their systems against onslaughts of users with errors that turn out to be faulty programming or bad data rather than system malfunctions.
4. Expect your computing service to be shut down at times due to routine maintenance and occasional breakdowns.
5. Expect computer service to be somewhat erratic due to varying user loads. When demand is heavy: in hands-on service, reservations are harder to make; in batch processing, turnaround times increase; and in time sharing, waiting time between messages increases. When the load is light, service to you as an individual user will be at par for your facility.
6. Expect a programming project to take longer than you think it will. Hence, begin writing and debugging the program as soon as possible, but not before a clear statement (usually in terms of a flow diagram) of what the computer is to do is at hand.
7. Set a level of excellence that will satisfy you and then stop polishing your program when that level is reached. A complex programming project will be finished only when you declare it finished. At that point, or at any point thereafter, expect that you will always want to change a few features when time and resources permit.
8. Carefully document all programs and data you wish to keep, otherwise you will later spend more time trying to discover what it is you have kept than it took to create it originally.
9. Always maintain test data for your programs, that, when run, will tell you whether the programs are doing what you want them to do.
10. Do not hesitate to ask for help, but ask someone who is designated to supply this service. One of the objectives this per-

son will have will be to eliminate inefficiencies in your program thereby reducing running time and hence cost to you.

11. Leave the equipment and other facilities you use as clean and attractive as you would like to find them when you arrive.
12. Keep informed of the rules users must follow. Rules change from time to time as a computer system evolves. Follow these rules exactly, since the staff operators will expect that you have done so when they run your programs for you. If you think the rules need improvement, take the matter up through proper channels, not with the staff operators who themselves must follow the existing rules.
13. Finally, remember a computer system is an expensive helper. For most users the computer system itself is not an object of primary interest or study. However, when you first learn to use a computer, you may experience an initial enchantment so compelling that you devote more time than justified by your project just putting the computer through its paces. Once you have passed this "first-love" phase, and abide by the above maxims, you may enjoy a rewarding life-long working partnership with computer systems.[1]

EXERCISES

5.1. Consider the man-computer simulation of the example game presented in Chapter 2. Plan to run this simulation for 30 participants. Assume there is a hands-on, open-shop computer that you can reserve for your purposes. Write a description of your whole simulation system. Be sure to state exactly what the computer does, what you do, and what the participants do. Use flow diagrams as a language where appropriate. Label the elements of Figure 5.1 where they appear in your description.

5.2. Do Exercise 5.1, except assume you have only closed-shop, batch-processing computer service. (If you have already done Exercise 5.1, state only the changes you would make in your simulation system description.)

5.3. Do Exercise 5.1, except assume you have available by telephone lines time-shared computing service from another city,

[1] After reviewing the above maxims for the author, one computing center director commented, "Amen!"

with only one remote terminal in your city. You can reserve this unit as needed. (If you have already done Exercise 5.1, state only the changes you would make in your simulation system description.)

5.4. Do Exercise 5.1, except assume you have available a room containing 31 remote terminals, each tied into the same time-sharing system within the same building in such a way that each can gain access to the same program and set of data residing in auxiliary storage. You can reserve this room as needed. (If you have already done Exercise 5.1, state only the changes you would make in your simulation system description.)

5.5. Consider the man-model simulation of the example game presented in Chapter 2. Plan to run this simulation for 15 pairs of participants using a computer to perform the administrative duties of determining a winner at each trail, keeping the indicated time paths and generated variables, and determining when to stop. Assume you have available a room containing 31 remote terminals, each tied into the same time-sharing system within the same building in such a way that each can gain access to the same program and set of data residing in auxiliary storage. You can reserve this room as needed. Write a description of your whole simulation system. Be sure to state what the computer does, what you do, and what the participants do. Use flow diagrams as a language where appropriate.

6

MAN-COMPUTER SIMULATION

Man-computer simulation differs from man-model simulation by the use of computers to execute models. This difference provides new opportunities for simulations containing man components. These opportunities appear in the complexity (and hence verisimilitude) of the model, in the amount and kinds of data obtained, and in the form and manner of presentation or receipt of stimuli and responses.

The use of computers in simulations containing live components also adds some constraints. Usually these are the need to design simulation systems to fit specific computer systems, to schedule sessions according to the availability of computer time, and to structure responses and stimuli to meet computer input-output requirements.

We shall continue to think in this chapter in terms of a stimulus-response (S-R) framework in which live entities receive stimuli and return responses. In man-model simulation, responses are returned to the administrator or to his assistants who then perform the operations called for by the model. In man-computer simulation, responses are returned ultimately to a computer. The computer then generates subsequent stimuli for presentation to participants.

Design of Man-Computer Simulations

The design of man-computer simulations requires all the ingredients and considerations given in Chapter 4 for man-model simulations, some of which were: time is simulated in discrete periods; pilot runs are recommended; live entities may be organized as interacting competitively, coordinatively, or both, or they may be independent; live entities may be single persons or teams; participants must be recruited, selected, briefed, and supervised; the model that represents part of the object system must be written; the administrator's role must be defined and his staff recruited and trained; and the whole simulation design must be documented.

In this chapter we shall discuss two more ingredients of both man-model and man-computer simulation. These are 1) the forms of stimuli, and 2) the forms of responses. We shall postpone discussion of programming man-computer simulation models until the next chapter when all-computer models may also be considered.

The general design of man-computer simulation is given in Figure 6.1. This design shows differences from man-model simulation only in the inclusion of computers in the stimulus-response segment. Although not shown in Figure 6.1, computers may also aid the administrator in briefing participants, cycling over time periods, assembling session data, and manipulating data for run and project reports.

Reduction of Interviewer, Experimenter, or Administration Bias

Human involvement in the administration of other human activities has always introduced added variability. Observations of identical phenomena by different persons are not always the same.

Briefings and answers to questions differ. The leadership of the experimenter or administrator, or the behavior of assistants, is seldom purely neutral. Where these variables influence the behavior of participants or the final data obtained, it is said in research that the results contain *interviewer* or *experimenter bias;* in education, *administration bias.*

The more computers take over contacts with participants, the more these biases are removed from simulations. When all stimuli come directly from the computer via an impersonal remote terminal operated only by the participant, when all responses are returned in a similar manner, and when briefings and answers to questions are given by the computer, a simulation may be said to be free of interviewer, experimenter, or administration bias. Of course, the bias that is eliminated is only the bias due to live contacts during briefings and sessions. Any biases built into the model by the designer or introduced by participant selection remain.

Flexibility

Flexibility is even more important in man-computer simulations than in man-model simulations.

Addition or removal of live entities during or between runs should be provided whenever possible. For example, in an online, real-time simulation laboratory, absence of participants should not shut down an entire session just because there are not enough persons to push all the keys or buttons the computer expects to be pushed.

Provisions for parallel runs (discussed on page 58 for man-model simulations) should also be considered when computers are used.

Built-in flexibility in man-computer simulations is sometimes crucial during actual runs because the amount of real time required to debug changes in the program may prevent adapting to unforeseen conditions. Simulation with computers lacks much of the immediate human adaptibility available when models are processed by live assistants.

Each computer simulation program will be written for a specific hardware-software system. However, the model and as much of the rest of the program as possible should be transferable to other computer systems. This means using a common language and standard input-output conventions. Computers are changed more frequently than expected and changeovers can be burdensome if this flexibility is not preplanned. Language and system flexibility is a necessity for educational simulations that are intended to run at many locations on different computers.

Period-to-Period Data Carryover

In designs providing sequential dependence, the current state of the model, as described by the values for its variables, must be carried forward to the next period. This carryover data is output from one period and input for the next. In man-model simulation, carryover data is usually stored as pencil marks on paper, as the positions of markers on a board, or as the number of tokens held (e.g. pennies in the example game).

In batch processing, carryover is easily handled by punching a carryover card deck. Only punched responses need be added to punched carryover to complete next period's input.

In some hands-on applications, the carryover data remains in main memory; in time-sharing, in auxiliary storage. In these last two cases, if more than one session is required for a simulation run, the carryover data must be stored outside the computer on cards or tape until the next session. In time-sharing systems, carryover data may sometimes remain in auxiliary storage for several days.

General Designs for Various Computer Systems

The kind of computer service available influences the design of a simulation. This may be unfortunate because the designer may consciously or unconsciously adapt his design to available service in a way that is antithetical to his original purposes. Sometimes he must compromise his purposes just to get his task done. For example, online, real-time, man-computer simulation may offer very advanced education or research, but be available only for few participants at a time due to the availability of remote terminals. If under these conditions participants come in large batches for only short periods, the designer may be forced to avoid the latest state of the art of simulation in order to accommodate all the participants.

The general design of man-computer simulations is given in Figure 6.1. Shown in Figures 6.2, 6.3, 6.4, and 6.5 are S-R segments that may be individually nested into Figure 6.1 for the following computer systems:

Figure 6.2. Hands-on, open-shop.
Figure 6.3. Closed-shop, batch-processing.
Figure 6.4. Online, real-time with a dedicated computer.
Figure 6.5. Time-sharing.

A design for running the man-computer session of our example game on a time-sharing system is shown in Figure 6.6. Here the administrator merely shows the participant how to use the remote terminal and then gets the simulation program "on the air." Thereafter, the program performs all administration directly through the terminal.

Definition of Time Period

Note that human behavior and responses come before stimuli in all these designs. Ordinarily one thinks of a stimulus coming before a response. The convention adopted in these diagrams depends on our definition of a time period in the whole simulation system.

Discrete time periods are represented in the diagrams as the cycle: human behavior—encoded responses—computer simulation—presentation of stimuli. The logic of the cycle is preserved when any one of the four phases is shown first. However, when the emphasis is on the computer portion of the simulation system, human behavior and encoded responses are shown first because they are input to the simulation program, and stimuli are shown last because they are the output of the program. For example, simulation of a year of operations in a management game requires 1) de-

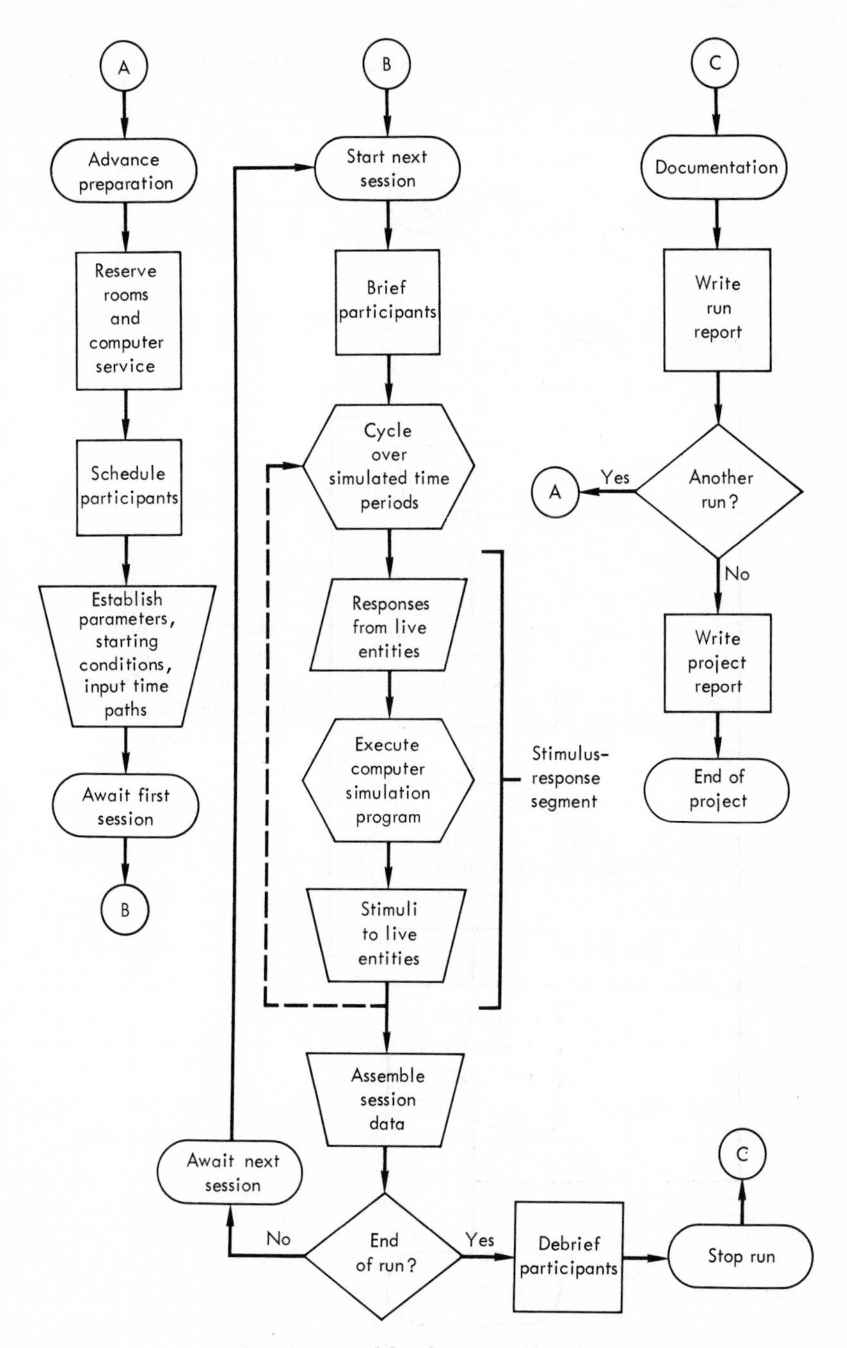

FIGURE 6.1. Man-Computer Simulation

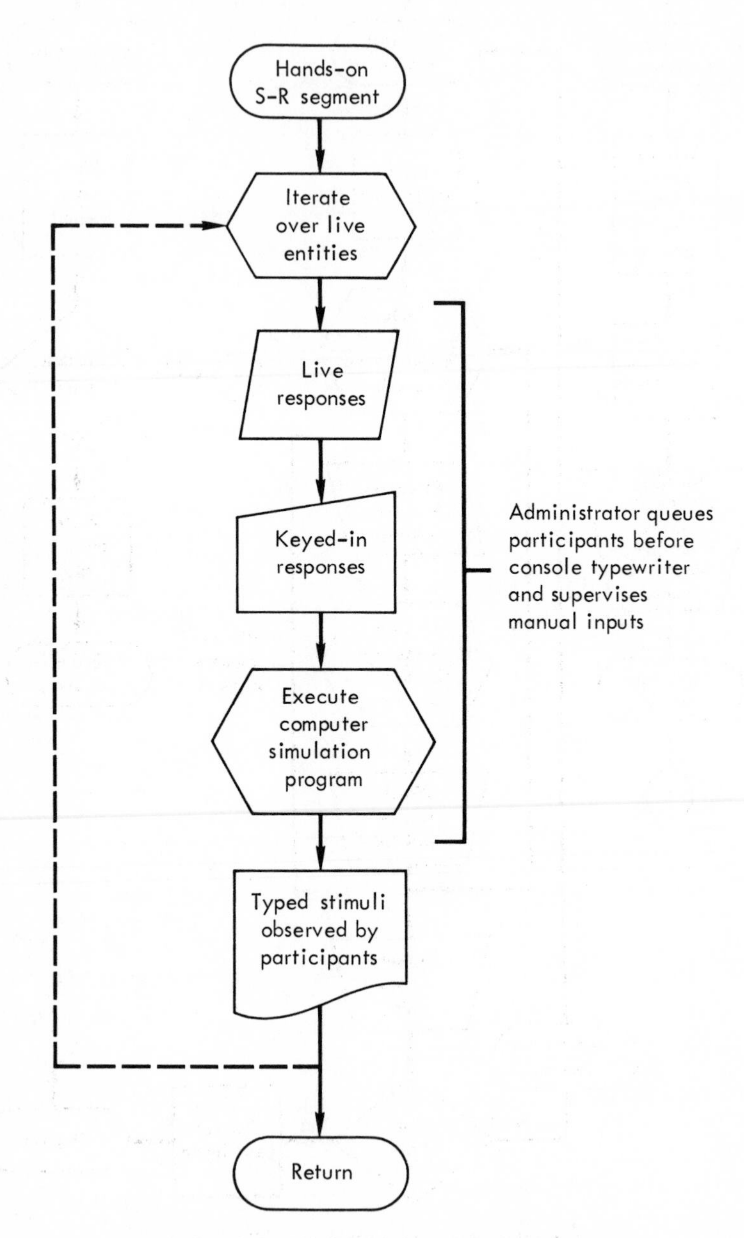

FIGURE 6.2. Hands-On, Open-Shop S-R Segment

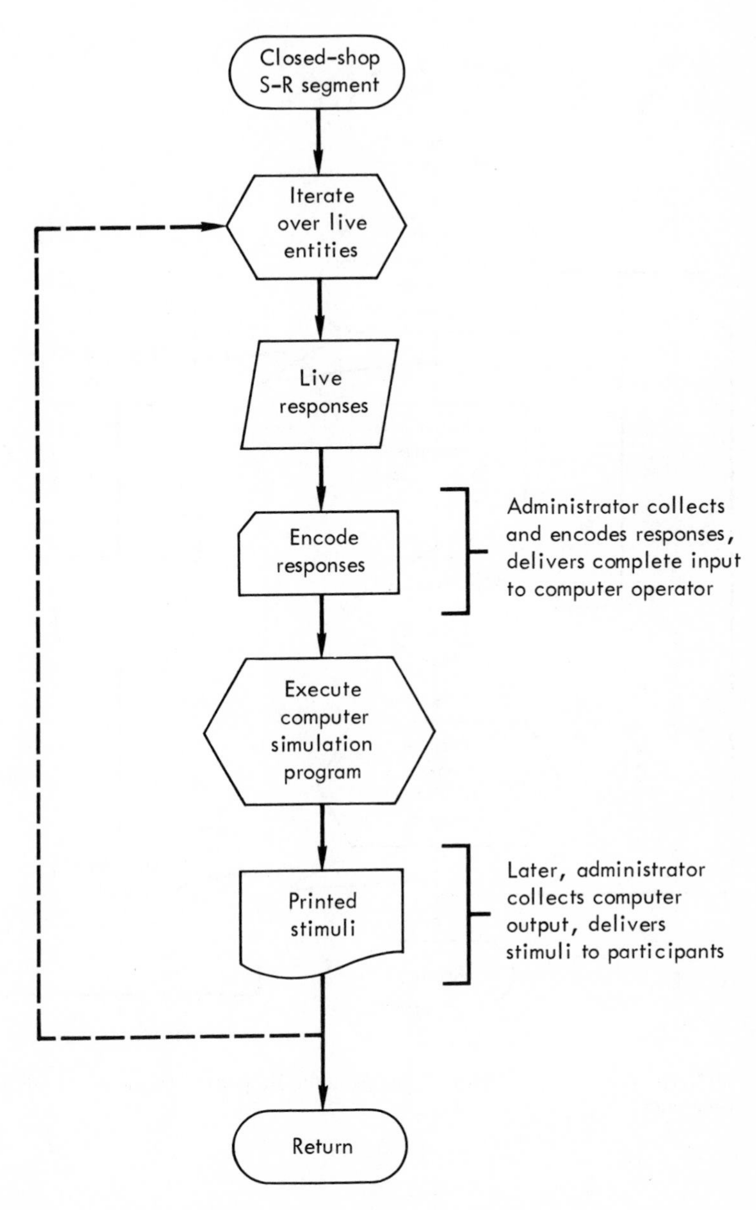

FIGURE 6.3. Closed-Shop, Batch-Processing Stimulus-Response Segment

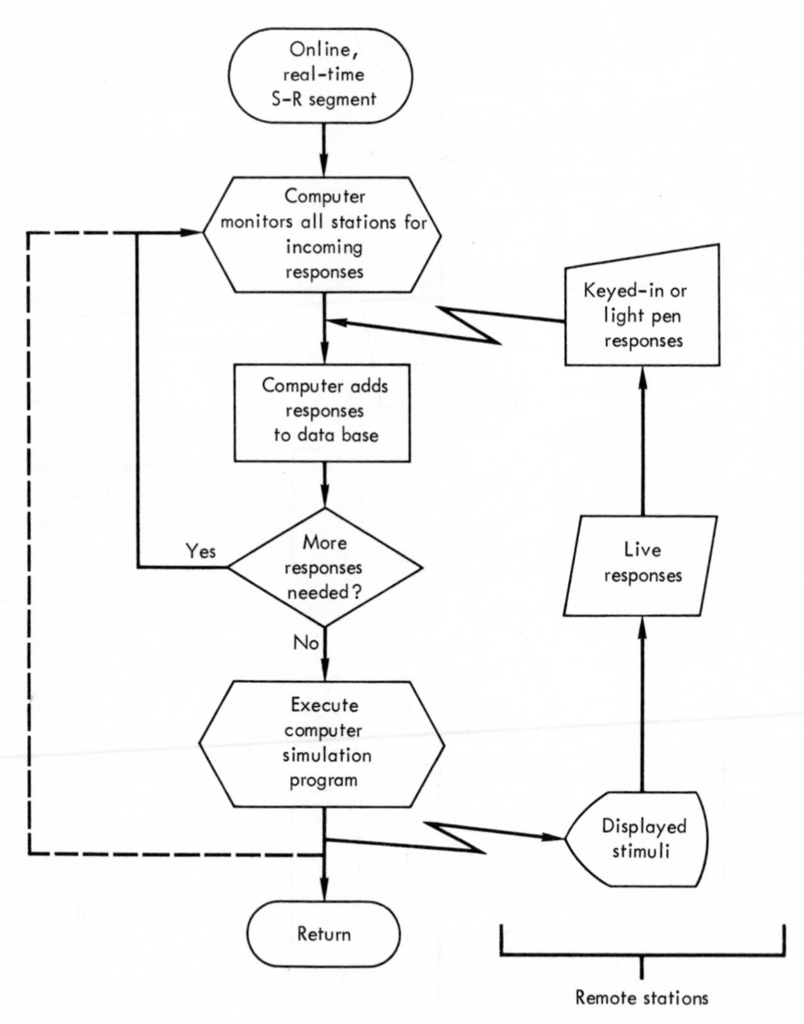

FIGURE 6.4. Online, Real-Time Stimulus-Response Segment (Dedicated Computer)

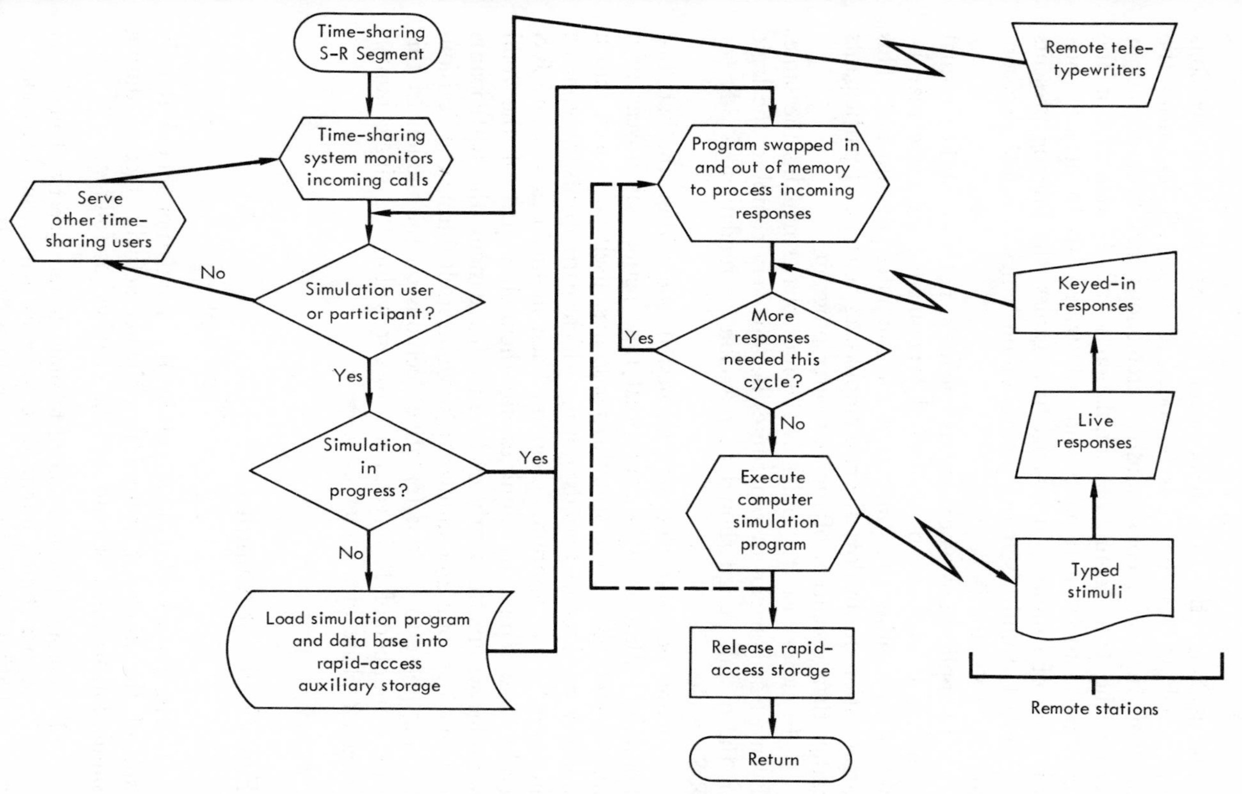

FIGURE 6.5. Time-Sharing S-R Segment

cision making (the human behavior), 2) writing down decisions and punching them on cards (encoded responses), 3) executing the program that represents industry interactions and firm operations during the year (computer simulation), and 4) delivery of the resulting financial statements and industry reports to players (presentation of stimuli). In management games, the briefing presents the initial stimuli (starting positions) on which the first decision-making behavior is based.

In applications where initial briefing is not also the first formal stimulus, as in many psychological experiments, the cycle may be shown with either the stimulus or the computer program first. This organization of our flow diagram emphasizes behavior of participants rather than execution of the computer program.

In either case, the order of the four phases remains the same. Even when behavior or procedures become very complex, the logic of the overall design should be reducible to fundamental S-R (or R-S) cycles.

Time period cycles may be nested one within another. For example, within a larger conceptual time period, say a year in a management game, the stimulus phase may itself be broken into a number of discrete periods during which participants may make inquiries of the computer program. Each inquiry is a separate R-S cycle. The replies to inquiries are themselves part of the total stimulus within the yearly cycle. In other designs, the participants may use the computer simulation program itself during the decision-making phase to try out tentative decisions on the model. In this case a number of R-S cycles may occur within the single human behavior phase of a larger cycle.

Four Types of Stimuli

The formal stimuli output by the simulation model is only one of the kinds of stimuli participants actually receive during simulation sessions. Stimuli may be classified as follows:

1. Formal stimuli intended to simulate the object system.
2. Other stimuli intended to influence participant behavior.
3. Stimuli intended to interrogate participants.
4. Irrelevant stimuli.

We shall discuss each type in turn. The discussion will include considerations pertinent to man-model simulation as well as man-computer simulation.

The total behavior of a participant during a simulation run depends not only on all the above types of stimuli but also on his background and on his activities between sessions. Some of these influences can be controlled as desired by proper selection and isolation of participants as discussed in Chapter 4.

Formal Stimuli to Simulate the Object System

During a simulation run, the only stimuli that directly represent the dynamic behavior of the object system are the outputs of the simulation model. Because the model and its input and output must be deliberately and unambiguously expressed, these stimuli are called *formal stimuli.* The burden of verisimilitude rests solely on the formal output of the model. Without verisimilitude, the exercise reduces to a numbers game or a mechanical interlude.

FORMS OF STIMULI

The output of the simulation model may be presented to live entities in a variety of ways: as pre-printed forms with blanks that are filled in each cycle by hand or by computer; as reports or exhibits completely generated by hand or by computer; as messages or exhibits projected on screens, written on chalkboards, typed on the console typewriter, typed on remote teleprinters, or displayed on CRT units; as announcements by the administrator or proctors; as pre-recorded textual or dramatic material over earphones or loudspeakers; as live telephone messages; as references to printed pages, or diagrams in possession of the participant; as actuation of special devices such as buzzers and lights; or as combinations of the preceding.

Figures 6.2, 6.3, 6.4, and 6.5 illustrate variations in formal stimuli presentation due to different kinds of computer service.

Within each means of presenting stimuli lie several alternatives. For example, teletypewriter messages may be sent as full sentences such as "I choose heads," as words such as "head" or "tail," or as symbols such as "H" or "T". Usually, the format of the stimuli will depend in part on the computer hardware, audio-visual equipment, or clerical services available. Yet within these constraints, wide variation is still possible. For example, should multiple-choice question formats be used? Should question-answer dynamics follow teaching-machine principles? Should questions be open-ended (despite processing and analysis difficulties)?

SENSITIVITY TO FORM OF STIMULI

The form of the stimuli and the manner of its presentation may affect participant behavior. In implementing our man-computer version of our example game, we might ask, "Is a head-tails pattern more quickly recognized when the stimuli are printed sentences? single words? symbols? vocal announcements? typed on teleprinters? displayed on CRT units? or projected on a large screen in a darkened room?" Definite answers to such questions have not been found. These issues are seldom considered. Most simulation designers use presentation means that are available or convenient under the assumption that variations in presentation mode will be irrelevant to the main purpose of the simulation.

REAL REWARDS

Differences in rewards do affect participant behavior, according to a mounting body of research and experience. Without a real take-home payoff, behavior tends more toward outguessing the model or seeking other intangible rewards than toward maximizing fictitious payoffs. As real payoffs (especially cash, although grade points in educational simulations are effective) are increased, behavior shifts toward maximizing individual payoffs. Of course, real payoffs should simulate payoffs in the object system just as in our example game a real penny served to simulate a $1,000 payoff. The size of cash payoffs depends on the resources available; unfortunately, research grants seldom include payoff money, yet this is one of the few variables in live simulations for which the behavioral effects are fairly well understood.

PRESENTATION SPEED

When stimuli are typed or displayed at slow speeds, as with remote teleprinters or CRT's that receive signals over ordinary telephone lines, participants soon learn the content of repeated messages and very quickly become bored and inattentive waiting for completion of typing of the message. To speed up the participant-computer dialogue, typed messages should be shortened as participants learn their meanings. To prevent shortened messages from becoming nonsense, an option to review the entire original message should be available.

With CRT displays at fast speeds, participants can visually select the parts of the display of interest to them. If participants are allowed to pace their own presentation speeds, they can quickly "flip" through a number of displays noting only information of interest.

MODIFICATIONS

Shortening of messages is one method of adapting stimuli to participants. Another method is simplification of the model. If the model is flexible, features can be eliminated to create a simpler version. For example, a management game that provides three products in two markets might be simplified to one product in one market. Or some decision elements in the game might be essentially "turned off" so that players make choices for fewer variables.

RECORDS OF FORMAL STIMULI

Records of formal stimuli should be preserved for several reasons. A record of stimuli may be needed for a later report. If subsequent replication of the behavioral conditions is desired, a complete record of formal stimuli must be available. Sometimes in computer-based runs it is easier to recreate the stimuli with the computer (by simulating the simulation) than to keep a current record. This technique and other data collection problems are discussed in Chapter 7.

Other Stimuli Intended to Influence Participants

The usefulness of simulations with man components derives from the whole simulation system, not the model alone. While model outputs simulate the object system, a variety of other stimuli condition participants' interactions with the model.

Where the model output refers directly to books, charts, or other materials in view or possession of the participant, these materials are considered part of the model's output. Where such materials relate only indirectly to model outputs, they are "other" stimuli. For example, a booklet of complete texts of abbreviated messages is part of the model, while background materials given during briefing are "other" stimuli.

BRIEFINGS

The initial briefing is singularly important because it sets the atmosphere for the entire simulation run. Although participants may learn further details after sessions are underway, it is fundamental to the notion of simulation that they have a basic understanding of the object system and the roles their behavior will represent before giving any formal responses. Also, it is axiomatic that participants learn enough of mechanical and procedural details at the initial briefing to perform their roles. Procedures should never overwhelm reaction to model outputs as representations of the object system. If a computer is used, a brief review of computers and the part they play in simulations may be given to overcome any latent mysticism about computers.

Initial briefings are usually presented in person by the administrator, but they may also be given in pre-session printed material or by teaching machine techniques at online remote stations, as in Figure 6.6.

Briefings may be simple or elaborate. They may use photographs, motion pictures, or recordings to portray factories, products, simulated persons, or even dramatic episodes. The first briefings may give the history to date and then the current status (initial formal stimuli) as participants take over the simulated object system. Organization charts or other materials given to participants may be colorful as well as complex. Wide variations are possible. For example, a researcher may portray a business firm during briefing in the form of graphs while an educator may represent it by its financial statements.

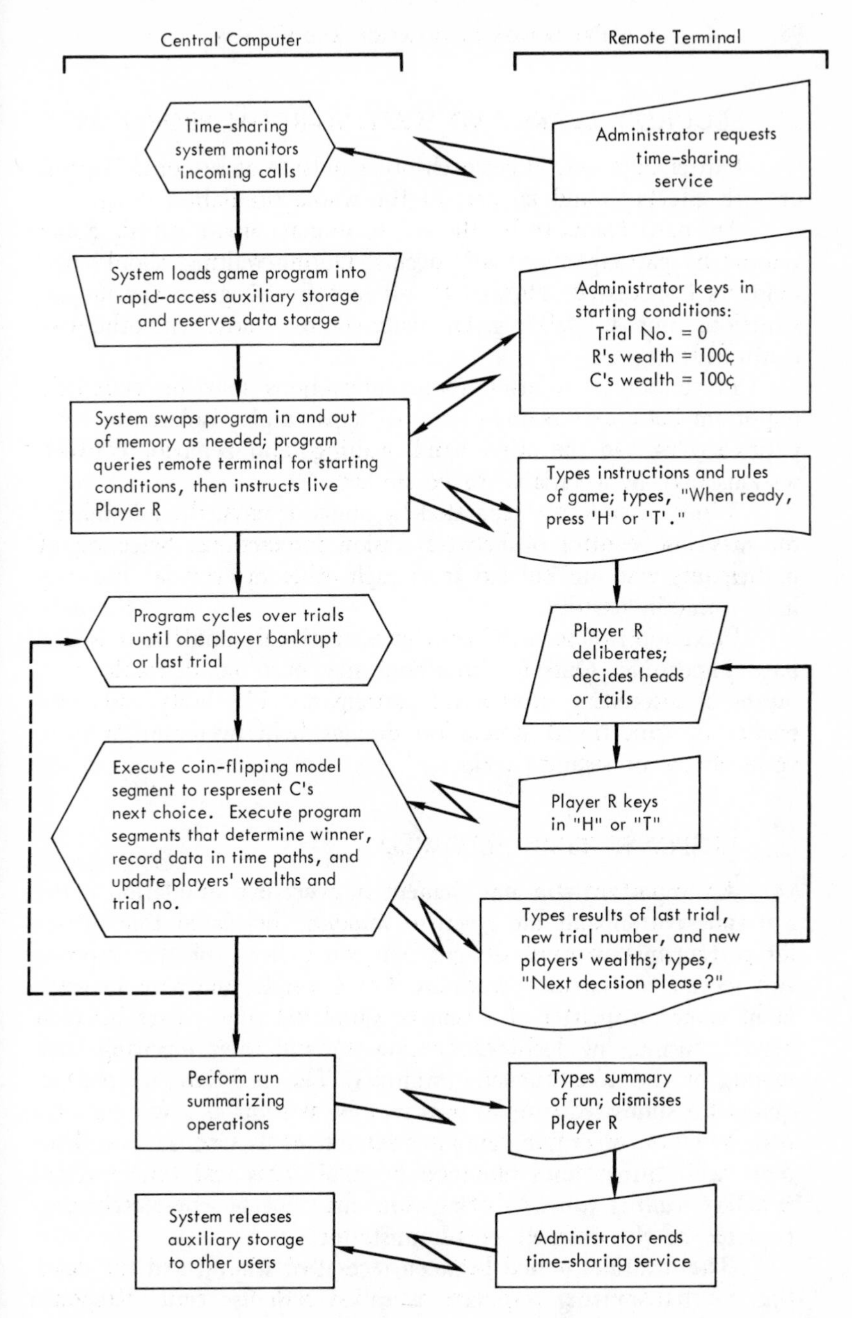

FIGURE 6.6. Time-Sharing Man-Computer Simulation of Example Game

SECURITY LEAKS AND WAITING-ROOM PROBLEMS

Participants may affect each other outside of sessions. Control of such effects should be part of the whole simulation design.

Are participants to be allowed to interact upon arrival? Some waiting by participants usually occurs. During waiting, should participants be isolated? allowed to interact freely? given specific instructions not to talk? given material to study? or otherwise controlled?

Prevention of waiting room interactions may be extremely important between sessions where participants play isolated or competitive roles. On the other hand, waiting-room behavior between sessions may be a variable to be studied.

When sessions are separated by hours or days, the administrator may lose control of between-session interactions. Selection of participants who are isolated from each other in everyday life may help maintain security.

Prevention of security leaks in educational simulations is perhaps impossible. Outside discussion may even be desirable as a means of interesting prospective participants. Flexibility and complexity in educational simulation designs help overcome any adverse effects of security leaks.

RESPONSE TIME ALLOWED

An important stimulus element in every live simulation is the real time participants are given to respond. Simulated time passes for participants in most designs between delivery of the response and receipt of the next stimulus. For example, in some management games, a quarter of a year of simulated time passes between players' turning in decisions (responses) and their receiving back reports of the consequences (stimuli). The real-time interval required for simulated time to pass may be as short as a few minutes with hands-on or online computer service, or as long as a week or more with turnarounds obtained by mail. This real time interval is called *waiting time*. Waiting time may not be completely controllable by the designer or administrator.

The real-time period between receipt of stimuli and the deadline for transmitting responses is called *response time*. Response time is a design controllable that affects participant behavior.

In general, the longer response time relative to the complexity of tasks, the more relaxed yet thorough participant reactions can be; the shorter response time, the more organized participant behavior must be. Response times that are either too short or too long engender snap judgments. If participants are too pressed for time, the mechanics of the simulation may overwhelm them and their responses may be meaningless. If they are undertasked during the response time, they may become bored and give insincere attention to response. Also, the longer the waiting time, the less participants remember. Very short waiting times may prevent needed rest. Pilot runs are invaluable for adjusting waiting and response times.

Stimuli Intended to Interrogate Participants

Stimuli intended to interrogate participants may be presented before, during, or after a simulation run. When personality or other tests are called for, they may be given before or during the initial briefing meeting. Interviews may be conducted at any time. Questionnaires may be requested at convenient times such as at rebriefings. These procedures are primarily data gathering techniques, but they may also provide clues about the object system or hints about participant roles and thus also function as "other" stimuli.

COMPUTER AS INTERROGATER

The manner of interrogation can also influence attitudes about the seriousness of the simulation. Computers not only help reduce interviewer, experimenter, or administrator bias, but time-sharing systems provide a unique opportunity for participant interrogation. The computer can ask questions via the same teletypewriter or CRT unit that delivers stimuli. The questions can be conditional on answers to earlier questions or on the output of the model. Most important, the computer can ask these questions at the moment of reaction and response, obtaining timely "on-the-spot" answers rather than later recall. Under present technology, replies to computer questions must be structured in advance by the simulation designer. Usually they take the form of multiple-choice questions.

PROTOCOLS

Another method of obtaining timely reports of participant behavior is to ask each participant to "think aloud" his reasons for doing what he is doing while he is doing it. Such a record is called a *protocol.* Protocols may be obtained by almost constant interviewing during sessions, or by asking participants to write down or tape-record their thoughts as they think them. Protocol information is open-ended; there is complete freedom of expression. A combination of highly structured computer questions and open-ended protocols is possible in online simulations. This is accomplished through computer-given instructions to write down, or speak into a microphone, additional "thoughts."

Irrelevant Stimuli

Many kinds of stimuli are assumed irrelevant to the purposes of the simulation because insufficient research exists on the sensitivity of live simulation results to such factors. Resources available usually prevent extra pilot sessions or repeated main runs to test whether selected stimuli are really irrelevant.

Some of the many variables in a live simulation environment that are presumed irrelevant are room temperature, relative humidity, degree of illumination, extraneous sounds, the general noise level (a significant factor in the choice of CRT units over teletypewriters when many units are located in one room), means of isolating participants (high screens, low screens, windowless rooms, grouping within a room, etc.). Of course, any factor presumed irrelevant within its normal ranges may be disruptive when extremes occur.

Four Types of Responses

Everything participants do or think during simulation sessions (or even between sessions) may be considered responses to the total stimuli presented. Not all of this behavior is relevant and useful.

Fortunately, the quantity of response obtained is controllable. Responses may be classified as follows:

1. Formal responses required by the model.
2. Other simulation role behavior.
3. Replies to interrogations.
4. Irrelevant behavior.

Again, we shall discuss each type in turn. In live simulation studies, we view participant behavior through the filter of simulation roles in a manner that ignores much of the total behavior. This is true of all of the three types of responses considered relevant.

Required Formal Responses

Formal responses required by the model usually take the form of numerical quantities or logical choices. The model as a set of operations cannot operate with open-ended or vague responses. In our example game in Figure 6.6, a formal response is pressing the "H" key or the "T" key. In management games, formal responses are sets of numbers representing decisions for those variables controllable by participants. In online competitive board games, formal responses are numbers and letters representing co-ordinates of moves to be made by tokens on a visually displayed board. In some experimental economics games, formal responses are merely prices written on slips of paper.

The amount and kinds of formal responses must be strictly determined when the simulation is designed. Data collection considerations can shape formal response design; for example, the need for highly structured replies when the computer acts as interrogator.

FORMS OF RESPONSES

Physically, formal responses may be returned by participants as ordinary handwriting or as coded marks on special paper forms that can be read by optical scanners; as vocal messages in person or by telephone; as punched cards or tape (special cards that can be punched by pencil are convenient); as keyed-in characters on console typewriters, keyboards, or teletypewriters; or as light pen signals. Handwritten responses may be encoded for computer input by the participants themselves; in this case, the administrator receives both the written responses and the punched cards or paper tape so he can verify the punching and also later check whether the computer read what the participants intended.

QUALITATIVE JUDGMENTS

A model may require qualitative judgments as formal input. These do not come directly from participants as responses but are decisions made by referees, a panel of judges, or the administrator following the observation of participant behavior or the reading of reports prepared by participants. The input is usually an index number. In computer processed models, it is necessary that neutral values for qualitative indexes be provided so the model can operate in the absence of qualitative judgments. Again, this points to the need for pre-planned flexibility in man-computer simulations; here, the capability to run with or without the qualitative judgment feature.

PARTICIPATION REQUIREMENT

For each cycle, the combination of the formal responses required by the model and any additional reports, meetings, interviews, or other activities required by the administrator for setting qualitative indexes is called the *participation requirement.*

Other Role Behavior

Participant behavior other than formal responses may be partially unpredictable at the time the simulation is designed. During sessions, it is controllable to some extent, hopefully to a degree that keeps the simulation within intended bounds.

ROLE-BEHAVIOR DATA

All nonformal role-related behavior is potential data for subsequent study and analysis. Some of it can be captured by means of the qualitative indexes mentioned above that become formal inputs to the model. Our present capacity for recording data of this nonformal type far exceeds our ability to do something useful with it later. Past simulation experience shows that massive amounts of data have been collected which were then never analyzed and used. Unless carefully planned in advance, the collected data from a simulation project may overwhelm the researcher in his effort to synthesize a report.

COMPLEXITY OF BEHAVIOR

Live behavior in simulations is not simple, even in such rudimentary simulations as our penny-matching example. Like behavior in real object systems, behavior in simulations is too complex for complete explanation. Only a few variables are selected for subsequent analysis. Fortunately, unlike object-system behavior, behavior in simulations can be manipulated and guided so that the effects of these variables can be partially isolated. Moreover, observational opportunities in simulations permit the study of nuances of human behavior that escape detection in real object systems.

RECORDING BEHAVIOR

Possibilities for recording nonformal role behavior in live simulations appear unlimited. Participants may be photographed with motion picture cameras, monitored on closed-circuit television, recorded on sound or video tapes, observed from special booths with one-way glass, or observed by stooges placed among them (who, you will recall, may also exhibit structured behavior as an experimental variable). In addition, role behavior may be traced by monitored telephone conversations, written protocols, collected scratch paper, or by the piles of messages and message copies that accumulate when all commmunication is by hand-written notes.

Replies to Interrogations

The forms of replies to interrogations are usually specified in the interrogation stimuli themselves; for example, booklet questionnaires are given with pencil-and-paper response sheets, online questioning by computer provides its own hardware mode of response, and interviewing obtains vocal responses usually recorded on sound tape. Except for open-ended interrogation, such as interviewing and "thinking-aloud" protocols, responses to interrogations can be more highly structured and hence are easier to obtain and handle as data than the nonformal aspects of role behavior discussed above. This possibility suggests that the behavior to be studied should be captured through formal responses or structural interrogations whenever possible.

Irrelevant Responses

Irrelevant behavior, by definition, is behavior judged not pertinent to the purposes of the simulation. We do not intentionally make records of irrelevant behavior. However, much data is ultimately relegated to this category due to inability to use it in some meaningful way. The economically optimal time to consider relevancy is when the simulation project is designed.

Check List for Live Simulations

We have not reviewed all the ingredients of live simulations: participants, the model, the administrator's role, stimuli, and responses. A prospective first user of live simulations may ask at this point, "Yes, but now just what do I do to get started? Which features should I use?"

Chapters 4, 5, and 6 present the prospective live simulation user with a catalog of alternatives. Many of these alternatives will not be applicable for particular projects, either because facilities are not available or because some alternatives are obviously inappropriate. Yet, many alternatives remain from which the user may select the final features of his simulation design. The variety of simulation applications is so wide that no one recipe can be supplied. Each user must make decisions for himself on the relevance, usefulness, and feasibility of these features for his project. Research and education has always been partly creative, and the design of simulations offers a fertile domain for creativity.

Creativity, however, always departs from an established structure. We present below a check list for use in both education and research when planning live simulation projects. This list is not a prescription. It serves only to prevent oversights in the early stages of planning. It is an established structure from which creativity may proceed.

1. *Object System.* What real (past, present, future) or hypothetical system are we interested in? Can we describe it in words? What do we know already? Have we reviewed existing theory for this kind of system?
2. *Purpose.* Why this object system? What are we trying to do? What will be different when the project is over?
3. *Initial Design.* Can we create a tentative dynamic model of the whole object system? Can we state this model in a flow diagram? What aspects do we ignore? What do we include? And at what levels of abstraction or aggregation?
4. *Simulation.* Is simulation necessary? Can the object system be studied by analysis, direct observation, field experiment, or case histories? If for research, must the simulation be live? Why not all-computer?
5. *Man-Component.* How do we break out the man-component from our first model? How is a live entity defined? Will there be more than one kind of live entity? Teams or individuals? Do they interact? If so, face to face, or through the model?
6. *Participants.* What background or training should participants have? Are such participants available? What compromises from ideal participants must be made? Will we pay them? When are they available? Do we supply travel and food? How many "spare" participants should we have?
7. *Model.* Is the distinction between live-entity roles and the model clear? Will it be clear to participants? Is our flow diagram of the simulation model unambiguous? Will we process it by hand or by computer? How soon can we start debugging? Have we made the model unduly complicated? Is it efficient? (See discussion of computer model construction in next chapter.)
8. *Administrator's Role.* Who will do the administration? Arrangements made for proctors and processing assistants? Administrator's duties outlined? How much counseling will be done by proctors during the run? Could another person be the administrator? Would the instructions be clear to him? When will the assistants be trained? What will be the costs of administration?
9. *Stimuli.* What physical form? What have other designers used in the past? Should we attempt to make stimuli look

like object system counterparts? How do we bridge any gaps between "real-thing" stimuli and what we shall present? Should a single stimulus be presented in segments, say pages or sequences of messages? Who receives which stimuli? Can participants formally request information? Will manuals or charts be helpful? What about participant isolation to prevent stimuli leakage? Any other security arrangements?

10. *Responses.* Exactly what formal responses will be required? Are these uniform over all live entities? What in addition will comprise the total participation requirement? Are the mechanical aspects of formal responses as simple as possible? Should a single response be received all at once or as a sequence of questions and answers? Do planned formal responses meet all the needs of the model? How do we relate for participants our formal responses to what live counterparts in the object system actually do?
11. *Data Collection.* What data from the stimuli and responses do we wish to save? Any additional formal interrogations? Do we want observations on the informal role behavior of participants? Through what media? Should we pretest data collection procedures?
12. *Physical Facilities.* What is available? Modifications needed? Computer service? Costs?
13. *Overview.* Real time required? Total costs? Number of runs? Proposal written?
14. *Pilot Run.* Length for sufficient check-out? Model processing trials? Participants for pilot runs? Any unexpected behavior or unusual results?
15. *Final Design.* Changes due to pilot run? Do time and cost estimates need revision? Recheck availabilities and commitments by others.
16. *Scheduling and Running.* At this point, the flow diagrams of the text take over check list functions.

EXERCISES

6.1. For the man-computer simulation of the example game of Chapter 2, design a simulation system, using flow diagrams, that will interrogate each subject about his "momentary" reasons for the choice he made at each trial. Remember that Player C chooses tails with probability two-thirds. (If you have previously designed or described a simulation system for this version of the example game, state only the changes you would make to meet the requirements of this exercise.)

6.2. Execute for one participant your design for Exercise 6.1. Do this on a computer if you can, otherwise execute your design as a man-model simulation with you as administrator performing the role of the computer. Do not exceed one hour of real run time. Prepare a presentation of the protocol obtained and an analysis based on the protocol of the behavior of the participant.

6.3. Write a report of the execution of Exercise 6.2 that classifies, describes, and discusses the stimuli and responses according to the scheme and considerations presented in this chapter.

6.4. Write a report of the execution of a man-computer simulation you have participated in or read about (but not one from Chapters 1-6 of this book) that classifies, describes, and discusses the stimuli and responses according to the scheme and considerations presented in this chapter.

6.5. Repeat your execution of Exercise 6.2, using the same time path of heads and tails by the computer, but varying some one stimulus or condition you previously thought was irrelevant. Make a judgment whether your original conclusion of irrelevancy was correct.

6.6. Repeat your execution of Exercise 6.2, using the same time path of heads and tails by the computer, but vary some one relevant stimulus or condition discussed in the chapter. Make a judgment about the effect of changing this condition.

6.7. Repeat Exercise 6.6, but for a different relevant stimulus or condition.

6.8. For the automobile repair bidding object system of Exercise 4.2, design a complete man-computer simulation in which the computer represents one of the three bidders. Try to give consideration to every aspect of man-component simulations discussed in Chapters 4 and 6.

7

ALL-COMPUTER SIMULATION

All-computer simulation is the last of the four general techniques categorized in this book for studying object systems. Recall that we study object systems for the purposes of describing them, explaining them, or predicting their future behavior. Of the four techniques, analysis and all-computer simulation have no live or man component. (Simulations without a live component are sometimes called *closed-loop* simulations; those with a live component, *open-loop* simulations.)

There is an important distinction between analysis and all-computer simulation. To clarify this difference, let us again use our simple example of the penny-matching game. So far in this book, we have not developed the version of this game that most of us have seen played, the one in which both live players make their decisions by physically flipping pennies. If both players actually flipped pennies to make their decisions, the trial-by-trial outcomes of the game and the ending results would form a process governed wholly by chance. In Chapter 8 on Monte Carlo simulation, we will show the analytical model of this process. Given particular inputs for this analytical model, it can tell us the frequency of occurrence of all possible ending numbers of wins for each player.

In the man-computer version of the example game in Chapter 2, a computer was programmed to simulate flipping a penny which had probability of tails equal to two-thirds. This computer program was used to represent Player C. For the all-computer version of Figure 3.3, we merely substituted a similar computer program for Player R. The computer then "thought through" the specified number of trials and gave the ending trial number, ending wealths, number of wins for each player, and a time path of the proportion of wins to date for Player R. If we allowed players unlimited wealth, repeated this all-computer simulation many times, and tallied the ending number of wins for each run, these tallies would also show the frequency of occurrence of possible ending numbers of wins for each player. We would have generated conclusions similar to those produced by analysis, but with a great deal more work and possibly less accuracy.

There is a moral here: simulation is never needed when analysis will suffice. You may remember the following paragraph from Chapter 3:

> Both analytical models and simulation models can be manipulated, but analytical models by definition are not experiments. Study of the model to reach conclusions about the object system can be attempted by analysis, by experiment, or by both methods. Many times analysis alone will serve our purposes, and analysis is usually easier, quicker, and less expensive than an experiment. At other times, analysis may yield incomplete results or the model may be too complex for direct mathematical attack; in these cases, experiments are the only way we can use our model to reach conclusions about the object system it represents.

Analysis is further distinguished from computer simulation as follows:

1. Analysis is usually a single-attack methodology with models that respresent either all possible instances of object system behavior or a single representative case. Simulation is a case-by-case technique that produces a specific number of individual

representations of an object system. (For this reason, simulation rather than analysis has become the accepted method for studying object system dynamics.)

2. Analysis produces conclusions directly from a model. Simulation generalizes from repeated applications of a model.
3. Simulation may use analytical models, but not *vice versa*. One form of simulation is repeated application of an analytical model but with different inputs for each application.

Historical Simulation

Simulations representing phenomena that have already occurred are called *historical simulations*. Because all-computer simulation has none of the run-time human variability found in man-model and man-computer simulation, it is ideal for this purpose. All-computer simulation can give identical results run after run (with the exception of those computer programs tied to outside-the-model processes such as a real-time clock). This perfect repeatability is a characteristic we desire when we wish to imitate an object system whose behavior is already determined by having happened.

Records of the past usually contain no uncertainties. There may be questions of what the record is about or what the data represent, but the written or spoken records themselves may be taken as given. This simply means that observations from the past have the same values each time we inspect them. Thus, unless something is wrong with the computer system, an all-computer simulation using past data should give us identical results run after run.

As a test of the validity of a simulation model, output from a run with given historical inputs can be compared with the recorded historical output of the object system itself. For example, a computer program might be written to represent the administrator's duties in man-model runs of the penny-matching game. This program could be executed on the historical sequences of heads and tails actually chosen by live players. The output of this computer program would be time paths of generated variables. We could compare the historical simulation outputs with the counterpart

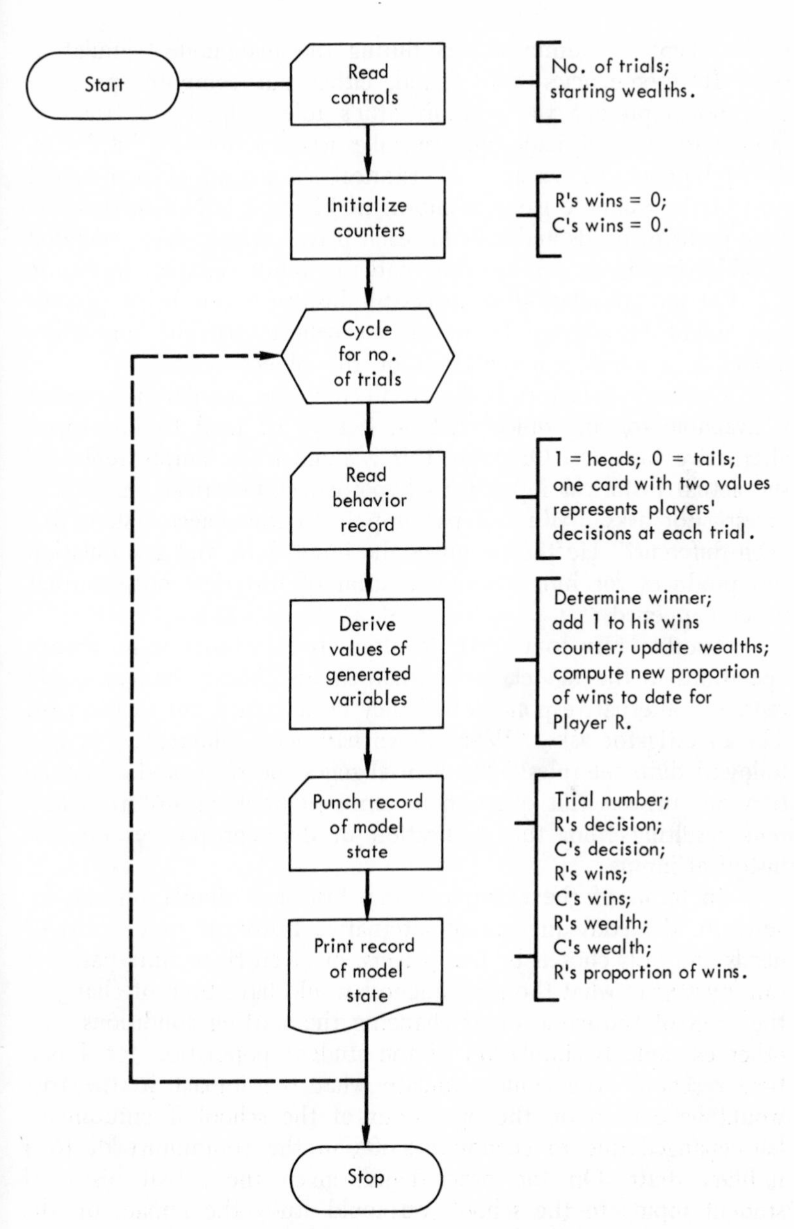

FIGURE 7.1. Batch-Processing Historical All-Computer Simulation

records kept by administrators during the man-model simulation runs. If discrepencies were found, either our computer program does not represent the administrator's role properly, or the administrator himself made some mistakes when performing his duties. A batch-processing version of historical simulation of man-model runs of the example game is shown in Figure 7.1. Notice that the time paths of heads and tails for each player, which were generated variables in Figure 3.3, are time paths of input variables in Figure 7.1. For any simulation, a successful historical run helps persons who might be affected by its output believe that the simulation model is a valid representation of the object system.

Once a satisfactory historical all-computer simulation program is available for any object system, it may be used to investigate alternative histories. One way this is done is by simply replacing the actual historical input with hypothetical historical input. The investigator asks, "What if past inputs to the object system had been different?" He then changes the input data, and a simulation run produces for him a representation of historical outputs that never happened.

Another way to investigate alternative histories is to change operations in the model, rather than inputs. These changes would represent alternative policies had they been carried out in the past. The investigator asks, "What if we had acted differently, or had followed different rules?" He then changes the rules, and a simulation run answers this question in terms of fictitious historical outputs resulting from the application of different policies on real historical inputs.

In terms of the example game, historical simulation can investigate the consequences of alternative historical time paths of heads and tails chosen by live players, or given these time paths, it can investigate what the consequences would have been of changing the rules of the game or of changing the starting conditions. Another example is simulation of the student population for a particular school. We could simulate what the impact in the past would have been on the operations of the school if enrollments had changed due to economic shifts in the community or to a military draft. On the other hand, given the actual historical student inputs to the school, we could study the impact on the student population and on school operations of changing, say, the rules for student suspension.

Future Simulation

A next step is to assert that future object system behavior can be represented by the historical model, or by variations of its inputs or operations.

INPUT VARIATIONS

By using hypothetical future data as input for an all-computer simulation of a historical object system, hypothetical future output behavior may be generated. If the future may occur in a number of possible different ways, represented by unique sets of input and output data, these can be generated by executing the computer simulation on each set of inputs.

MODEL VARIATIONS

If the future object system will itself differ from the historical system, changes may be made in the simulation model to represent the object system of the future. If several different future object systems are possible, the behavior of each may be investigated by runs with corresponding model changes. For comparability, these runs should use identical hypothetical future input data.

BOTH INPUT AND MODEL VARIATIONS

To study the influence of input variations on any given model of a future object system, the model may be executed on different sets of hypothetical future data. The combined effects of input variations and model variations may be studied by running changed data on a changed object system model. However, to avoid confounding of effects, it is necesary to run either changed data on a given model or a changed model on given data.

VARYING CONTROLLABLES AND UNCONTROLLABLES

Either or both the inputs and the operations of the object system may be under managerial control. More often only some inputs and some operations are controllable while the remaining are either constant or vary uncontrollably. Uncontrollables may be predictable to some degree. Managers and administrators in business, government, and education can use future simulation to study the impact of future decisions on object systems that contain uncontrollable variations. By this means, they can evaluate alternative adaptive adjustments to those things that they cannot control. Earlier, we wished to avoid confounding effects of changed input with changed model operations. Similarly, it is desirable to avoid confounding the effect of changes in controllables with the effects of uncontrollables. Hence, a logical sequence of simulation runs is:

1. Historical simulation.
2. Predicted uncontrollables run with the same controllables.
3. Changed controllables run with the historical values of the uncontrollables.
4. Adaptive changes in the controllables run with predicted uncontrollables.

Following this procedure, managers can investigate the consequences of their estimates of future uncontrollables, the consequences of alternative managerial responses to these changed conditions, and also the effect these changed policies would have had in the past. See Figure 7.2 for an online version of this procedure. The computer programs that provide these services are sometimes called "what if" programs. "What if" studies may lead to decisions that are actually implemented in the object system. Subsequently, a historical simulation would then aid understanding of what actually happened. The historical simulation could then be run on newly predicted uncontrollables and on changed controllables, and so on. Repetition of these cycles is *management by simulation.*

REAL-TIME CONTROL

When adaptive changes have significant immediate effects, the system is called process control. Inputs to the control system rapidly cause changes in object system controllables which in turn cause new object system outputs that become new inputs for the control system. The classic example of a control system is the thermostat. The special metal inside a thermostat moves with room temperature and causes a heat exchanger to be turned on or off. The action of the heat exchanger heats or cools the air temperature

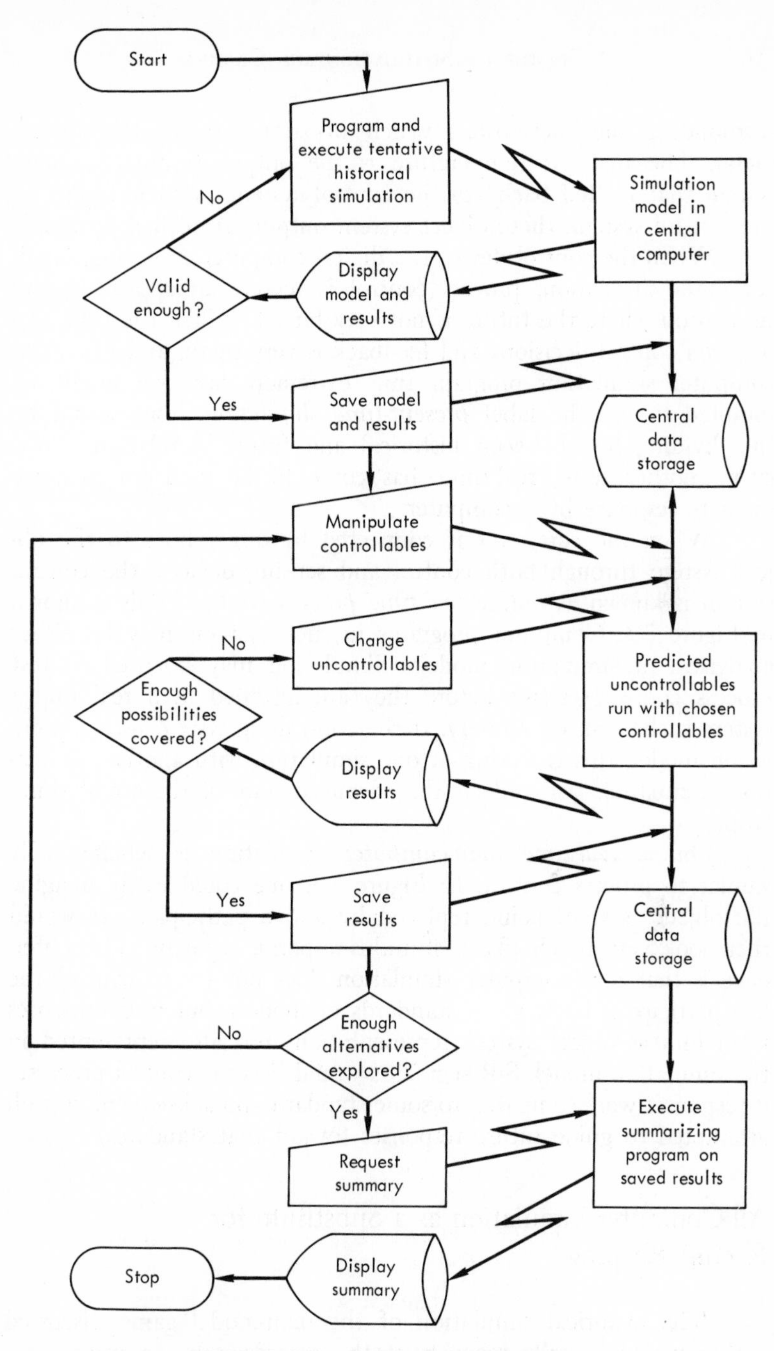

FIGURE 7.2. Online Future Simulation

surrounding the thermostat, which in turn actuates the special metal. The new air temperature is the output from the object system that is "fed back" to the control system. As new inputs to the control system, these object system outputs are called *feedback*.

When the control device is a digital computer executing an all-computer simulation, process control is merely an applied future simulation where the future is not very far off. When the time gap between control decisions and feedback is very small, and when the computer simulation program runs extremely fast, we might be tempted to use the label *present-time simulation*. This would be the dividing line between historical and future simulation. However, another term, real-time, has come to be used for near-immediate response by a computer.

When the computer is connected by direct lines to the object system through both control and sensing devices, the control system is known as *online real-time process control*. This is shown in Figure 7.3. Computer programs in such systems may be either analytical or simulation models. Simulation may be used to test process control systems before they are installed in a real object system such as an oil refinery. If the controlling system uses a simulation model, this is testing of one simulation with another simulation, a situation that reflects the advanced state of the art of simulation.

Online real-time man-computer simulation is schematically similar to process control. In Figure 7.3, one could easily imagine the object system being replaced by a live participant. It would then look very much like a stimulus-response segment. The difference is that man-computer simulation does not try to control the live participant to fit given standards or models, but uses responses from him as object system representations not also represented in the simulation model. S-R segments would become control processes if responses were compared to some standard and subsequent stimuli attempted to guide future responses toward that standard.

All-Computer Simulation as a Substitute for Record Keeping

The historical simulation of the man-model game discussed earlier would literally reconstruct the records a live administrator

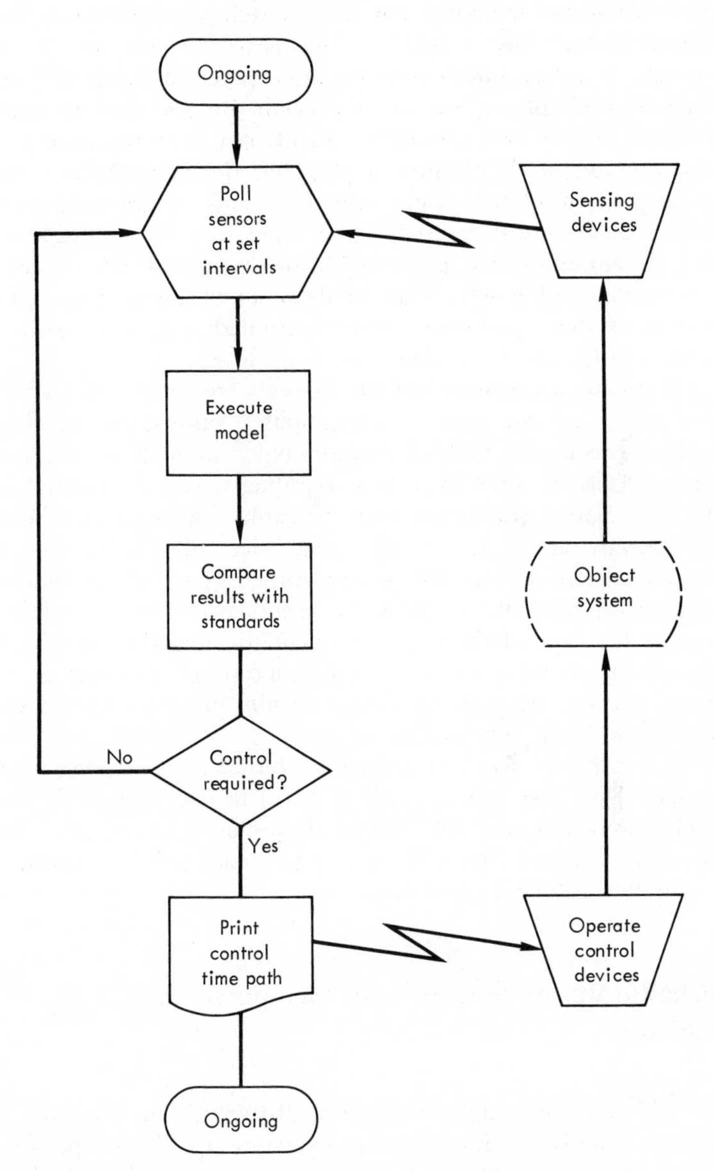

FIGURE 7.3. Online Real-Time Process Control

kept. It would not reproduce any records each player may have kept for himself, but these would not be needed to simulate the administrator's duties. Given only the time paths of heads and tails chosen by each player, we could generate by historical computer simulation all the observations we might have recorded during the actual experiment. With such a program, the administrator need only keep the heads-tails time paths. For many experiments, reconstruction by historical computer simulation may be a considerable saving in administrative labor and expense, and it may even be more accurate. Moreover, kinds of data not otherwise recorded at the time of the experiment can be generated; e.g., time paths of players' wealths or of individual wins and losses.

For complex man-model experiments, reconstruction of the experimental environment by all-computer simulation can be a valuable investigative tool. In man-computer simulations, the computer may also be used as a data recording device during the live run. This enables simulation users to capture a larger number of behavior variables. Then, using these "elementary" observations, users can derive by an after-the-experiment historical simulation generated variables that would be too cumbersome or impossible to record at the time of the live experiment. In fact, this may be the only way of studying inside-the-head conceptual processes of live subjects because attempts to observe mental processes in real time may destroy those very processes.

A possibility barely explored in business, government, and education is to derive later much of the data now kept daily. This would require only current records of specified "elementary" data —all other required "records" would be generated by subsequent all-computer historical simulation.

Stochastic Versus Nonstochastic Computer Simulation

The penny-flipping game used to distinguish analysis from all-computer simulation involved the operation of chance processes. Models of object system chance processes are called *stochastic processes.* All-computer simulation of the example game used stochastic processes at each trial by first determining the side appearing for one player and then the side appearing for the other player. These "decisions by chance" were simulated by a penny-flipping computer program segment that received as a parameter the appropriate probability for each coin represented. When this program is run, what sides are exposed by each simulated player differs trial to trial. If the program is run several times, individual time paths of heads and tails differ from run to run. Such repetitive execution of an all-computer simulation model containing stochastic processes is commonly called Monte Carlo simulation, which will be covered in Chapter 8.

To obtain all-computer outputs that differ from trial to trial, different inputs to the model are required each trial. Only if the computer is malfunctioning will outputs from a digital computer program differ when identical inputs are given to that program.

An all-computer simulation model, we should recall, is only a segment of a whole computer program. Inputs to the computer model itself from the whole computer program (which may have generated them or read them as inputs) may be of two kinds:

1. On the one hand, there are the starting conditions, parameters, and time paths of variable inputs that have already been discussed.
2. On the other hand, there are special inputs to the computer model which are required to produce stochastic variability trial after trial. These special inputs are *random numbers.* Each time an event from a stochastic process is needed, one or more random numbers are used up. Random numbers have properties that we will discuss in the next chapter.

In order to distinguish between stochastic and nonstochastic simulation, we must be able to distinguish between the special inputs that are random numbers needed by computer models of stochastic processes and all the other kinds of inputs to a computer simulation model. The reason for this distinction is that we want to think of random numbers as being inherent in the operations of the stochastic processes and not an input or operation that simulation designers can fashion to their purposes. That is, we want to exclude random numbers from our list of model inputs, although, logically and in a final analysis, they can only be classed as model inputs because segments of the model need them to produce stochastic events.

Random numbers may be read one by one from an actual list stored on cards, tape, or in main memory. In terms that we have developed in this book, random numbers form a time path, sometimes called a tract of random numbers. Rather than being stored as a list, random numbers are more often generated one by one as needed by a special program segment, called a *random number generator*, external to the model but internal to the whole program. In either case, the resulting list of numbers itself is fixed and definite and is not a stochastic variable, although its purpose is to create stochastic variation in the model.

Let us forget for the moment that random numbers are really inputs to computer models to make stochastic process simulators work. With the random number "issue" out of the way, we shall distinguish between stochastic and nonstochastic simulation as follows:

1. *Simulation* is a case-by-case method for studying object systems. Each case may be either a single trial or an entire run. In either view, outputs may differ trial to trial and run to run.
2. A *stochastic simulation* is one in which differing outputs trial to trial can be obtained without changing the inputs (ignoring random numbers as inputs). Specifically, this means that identical parameters, starting conditions, and input time path values produce varying outputs trial to trial and run to run.
3. A *nonstochastic simulation* is one in which the inputs or the model must be changed to obtain changed outputs. This means that identical model operations, parameters, starting conditions, and input time path values will produce identical outputs run to run. With a given model, the only way to obtain different outputs trial to trial in nonstochastic simulation is to change time path inputs.

Historical simulation is an example of nonstochastic simulation. In our earlier illustration, variations to represent hypothetical histories were obtainable by varying input time paths or by changing operations of the model itself. Parameters and starting conditions could also have been changed to obtain changed outputs. Man-computer and all-computer simulation of the penny-flipping game is stochastic simulation since Player C chooses tails with a probability of two-thirds by a computerized stochastic process that represents the underlying chance process of flipping a real bent penny.

Nonstochastic simulations can be made stochastic by adding models of chance processes. Stochastic simulations can be made nonstochastic by "fixing" the stochastic processes so that they generate constant values.

An analogous distinction holds for analytical models. These can be *deterministic* or *probabilistic*. However, any analytical model is like a formula. Plug in its inputs, and the outputs are always the same. These outputs may describe stochastic processes, but the outputs are always identical with identical inputs. Both deterministic and probabilistic analytical models may be written in the form of computer simulation models and used for nonstochastic simulation. The fact that a probabilistic analytical model may be used in a nonstochastic simulation may appear to be a contradiction, but if we must vary inputs to obtain varying outputs from probabilistic analytical models, we still have nonstochastic computer simulation. Stochastic computer simulation does not occur until a computerized stochastic process, which indeed may be based on a probabilistic analytical model, is used to generate values that may differ from trial to trial regardless of changes in or the constancy of model inputs.

Two examples of nonstochastic simulation are 1) representation of underground water flows by repeated applications of formulas, and 2) simulation of a sequence of financial statements for a corporation. Either of these simulations may be made stochastic by adding models of chance processes: in the case of underground water flows, the uncertainty of weather on the earth's surface; and in the case of financial statements, the uncertainties of the marketplace.

Computer Simulation Programs

The way a simulation designer views an object system will depend on his own individual past training and experiences, on the language he intends to express his model in, and on whether a man component is to be included. Language types and the influence of each on modeling of object systems will be covered in Chapter 9; general computer programming considerations for all-computer and man-computer simulations are discussed in this chapter.

GUIDANCE OF PURPOSE

The purposes for which a computer simulation model is being constructed should dominate the writing of the program. Considerations of the available computer system and the constraints imposed by the computer language and the user's own computer experience often exert strong influences on the design of the model. When this happens, purposes become reversed. Sometimes it seems as if the object system and the task of simulating it are serving the computer system by providing an interesting thing for it to do.

Users, particularly new users, are warned not to allow the satisfaction of putting the computer system through its paces to influence the conceptual formulation of the simulation model. Obviously, the computer system, the languages, and the user's skills impose constraints, but these constraints should not impinge upon the user's conceptualization of the object system. Computer systems should extend the mind of man, not mold it. Moreover, excess involvement with the technicalities of computer systems tends to drive a simulation project off its track, and the result is huge amounts of output, lengthy and complicated computer programs, and no conclusions.

One way to keep a project focused on its purposes is to construct the simplest model conceivable, program it, run it, and draw conclusions however gross—and do this before proceeding to a finer model that is more representative of the object system.

PROGRAM AND MODEL SEGMENTS

Previously, we talked of segments of simulation models. We dealt at length with stimulus-response and interaction segments. There are also input segments, output segments, data storage segments, monitoring segments, and so on. These are logical parts of a whole simulation system.

Since computer programs are operations capable of repeating themselves, they are often called *routines*. When the task of writing a program is subdivided, the separate segments may be called *subroutines*. Subroutines are separate parts of a computer program that may be called into action when needed. Subroutines may call upon other subroutines. Small segments built into the main program at particular places rather than called upon are sometimes called *subprograms*. These are technical matters of computer programming specific to particular computer languages and systems.

For generality and convenience, we shall restrict ourselves to the idea of program segments. One of these must be a main or control segment. We do not imply that segments are necessarily organized as subprograms or subroutines in the sense of any one computer language or system. Program segments are sometimes called program *modules.*

INPUT SEGMENTS

Each computer program must at first be loaded or read into main memory (primary storage). The procedures for doing this are part of the computer system and are usually different at each installation. Thereafter, control is maintained by the user's program until relinquished back to the computer's operating system.

The first inputs on initiation of a user's simulation program are the control features designed by the user. Usually, these establish the number of runs and the number of trials per run. Then for each run, starting conditions and parameters are also read in. Input time paths may be loaded all at once at the time of initial input, or values from time paths in auxiliary storage, or on cards or tape, may be read at each cycle as needed.

In online man-computer simulation, special software features are provided by the computer system to monitor the remote terminals. Usually, this is a polling procedure that sequentially asks each terminal if a message is ready. Once the input is sent to primary data storage, it is available to the computer simulation model for the next cycle.

INDIVIDUAL S-R SEGMENTS

In man-computer simulation, stimulus-response segments may simply output data generated by interaction segments and then input responses. On the other hand, S-R segments may be elaborate simulation models themselves. When the segment takes over, it either generates the values of appropriate stimulus variables or obtains them from data storage, displays them to the live participant, obtains his response, and stores the values of the response in the appropriate locations in data storage. These values are then available to any interaction segments and to the S-R segment itself at a later cycle.

Complex responses from live participants can be obtained in a conversational or question-and-answer mode. These S-R segments can themselves be very large programs. This mode of obtaining responses from live participants has been most highly developed in the field of computer assisted instruction (CAI). Usually, the next message sent to a live CAI participant depends upon his previous responses, frequently just on his last response. CAI methods provide a highly interactive technique for dealing with participants in man-computer simulations. Used as an S-R segment, a CAI program may require during one simulation cycle a large number of man-computer conversational interactions, the number depending on the messages returned by the live participant. Not until the conversation is completed is a single response generated for delivery to the simulation model. Part of this conversation may be stimuli intended to interrogate subjects as discussed in Chapter 6. Responses to these interrogations must then be stored as data for later study.

In all-computer simulation, the output and input operations of S-R segments are not needed. Segments can access the required data directly. In creating an all-computer simulation of a man-computer simulation, the individual S-R output-input features must be replaced by all-computer models of the former live participants. The models receive stimuli from and deliver responses to data storage just as a man-computer S-R segment would.

INTERACTION SEGMENTS

Interaction segments require data that may have been generated or input in past cycles or input in the current cycle. To be available to interaction segments, these values must be in the computer's main memory (primary storage). This is a problem of data storage and management. The interaction segment needs merely go to this storage in computer memory for the values it requires. It then places the new values it generates into appropriate positions in data storage. These data replacements may or may not change the very values obtained earlier.

Multiple segments talk to each other through data storage. When data is not permanently stored within a computer's primary storage, say for reasons of bulk, the technique of communicating among segments by means of data storage may become burdensome, depending on the type of auxiliary storage device. In man-computer simulation, inputs from remote terminals go to data storage where they are called out as needed by interaction segments, but not before a check is run to see that all the inputs needed from all terminals are available. This procedure is illustrated in Figure 7.4

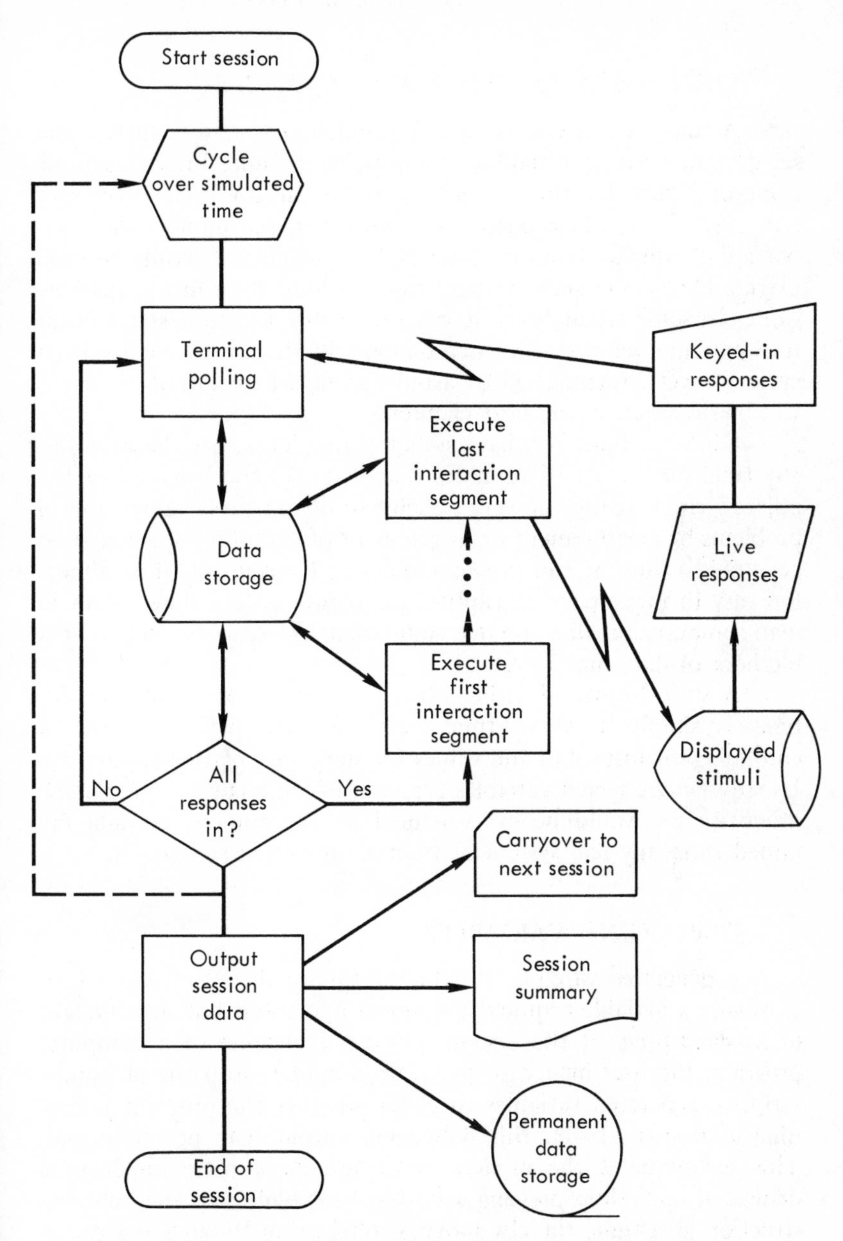

FIGURE 7.4. Data Management in Man-Computer Simulation

SYSTEM STATES AND STATE HISTORIES

A state of a whole computer simulation system is merely the set of values for its variables at any point in time. In the example game of Figure 3.3, the state of the system at any cycle is the current side of his coin exposed by each player, the number and proportion of wins to date for Player R, and the current wealth of each player. The system state changes from cycle to cycle in the example game. In some simulations, it is possible that identical states occur for some number of cycles when temporarily there is no change from cycle to cycle. (A method to avoid cycling when nothing happens will be discussed in the next chapter.)

Object systems also have system states. These are the values at any time for the set of variables that describes the behavior of the object system. It may be very difficult to obtain these values due to problems of measurement or to problems of capturing object system generated values at the proper moments. By contrast, it is almost too easy to measure or to capture simulation system state values in man-computer or all-computer simulations, so easy, in fact, that a plethora of data may result.

A state history of either object systems or of simulation systems is simply a set of time paths. A time path is merely a chronological history of the values for some one particular variable. Usually, only selected variables are used to represent a system state, otherwise we would be overwhelmed by the amount of data obtained from any real system or from a simulation system.

DEBUGGING VARIABLES

A generated variable for which a time path record is not kept is usually a variable required for model operations but not thought of as descriptive of the system. During debugging of a computer program, the user may wish to create output time paths of nondescriptive generated variables to check whether the program is running as desired. Later, this debugging output may be eliminated. This technique is the modern substitute for a programmer in a dedicated open-shop pressing a button to execute one computer instruction at a time, thereby moving step-by-step through the execution of his program to see that it runs as he wishes.

DATA STORAGE AND SEQUENTIAL DEPENDENCE

Data may be stored temporarily in many ways in a computer system. Primary storage is temporary because the data records kept in it disappear when the next user's program is serviced by the hardware system. More permanent storage is available on magnetic disks and tapes, and on punched cards. All of these may also be used for temporary storage. Cards that are punched during one run usually are not read again during the same run. Printed records and CRT displays are not considered data storage because they are not in machine-readable form, although they are definitely transfers of data to users and live participants.

The computer could conceivably absorb into primary storage all of the data requirements and all of the program segments needed for a single simulation run. These would be stored in the primary storage remaining after necessary dedication to the computer's operating system had been made. This whole complex of operating systems, programs, and data could then feed data items as needed to each segment of the computer model. It would also store all model outputs in primary storage until the end of the run when the entire contents of data storage could be printed for inspection and written onto cards or tape for subsequent analysis. However, this is not practical due to the limited size and great expense of primary storage. Moreover, in man-computer simulation, all the response data from participants is not available to be stored in the computer at the beginning of the run.

The simulation model may require at any cycle values of variables input or generated in past cycles. In this case, the current output of the simulation model depends on past data. If, as the simulation model is moved through time generating a sequence of system states, each state evolves wholly or in part out of the model's past behavior, we say that the simulation exhibits *sequential dependence.*

Sequential dependence is rather obviously a characteristic of many real systems. However, in the example game, we assumed *sequential independence* of the sides of the coin that may appear, i.e., which side appears does not depend on which sides appeared in the past. But for the other state values in the example game, sequential dependence does appear. Current values for players' wealths and for the proportions of wins by Player R are sequentially dependent on past values generated by the simulation model.

Sequential dependence implies that past data must be available for use by the model as needed. This availability must be in primary storage. Thus, these past values must either be retained in primary storage or be retrievable from auxiliary storage. Obtaining a value from either primary storage or auxiliary storage is called *accessing*. We say that the model can access past history in primary storage, that the program can access past history in auxiliary data storage, and that the computer can access whole programs or program segments (pages), either for the user's program or for its own operating system, from auxiliary "systems" storage.

Most problems of data management in simulation systems involve making the data on which the model is sequentially dependent available in primary storage. This must be accomplished in the primary storage left over after both the operating systems and the program (or its current segments) are loaded. These relationships are illustrated in Figure 7.5. Fortunately for the user, the computer system through the language provided performs many of these data management steps. However, the user should understand the organization of primary storage and the limitations implied by Figure 7.5 The larger the operating system in main memory, the less storage for programs and data. The more program segments in main memory, the less data, and *vice versa*.

INTERMEDIATE OUTPUT

Due to limitations of primary storage, an input time path may be read in value-by-value as needed in each cycle. In this way, only primary storage for the new value is required. If a separate record of this input is needed, this value may be output at each cycle. Also values obtained from live participants or values generated by the model at each cycle may be immediately output. Such within-run output is called *intermediate output*. Debugging variables may be easily handled as intermediate output. The purpose of intermediate output is to create output time paths without using up primary storage for more than current values or values required by sequential dependence. If displayed while running, intermediate output gives the effect of seeing the computer simulation program in action.

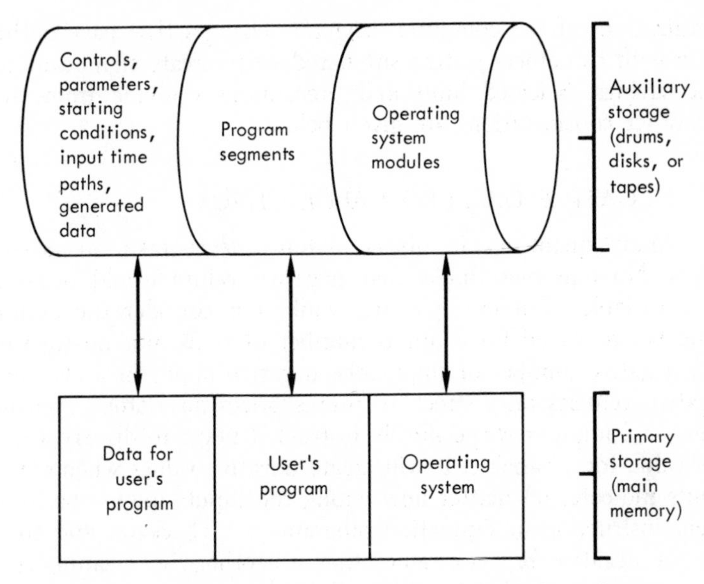

FIGURE 7.5. Organization of Primary and Auxiliary Storage

FINAL OUTPUT SEGMENTS

At the end of a specified run, the final output segment takes over. It may display for visual inspection and may write in machine-readable form any time paths not yet output, ending values for selected generated variables, and any values derived from final output such as averages and measures of variation. Derived values can also be generated by a separate computer analysis of readable intermediate and final output.

Hints and Precautions for Simulation Programming

Model builders sometimes have difficulty realizing that the computer cannot "think" of characteristics or conditions in object systems that are simply obvious or are taken for granted. Unless simulation programmers learn to attend to the obvious and to the assumed, strange things may happen. On the other hand, certain reorganizations of the computer program—changes that have nothing to do with the object system simulated—may create significant run-time savings. Selected hints and precautions evolving from these and other considerations are given below.

NEGATIVE OR ZERO VALUE CHECK

Many quantities in object systems never take on negative values. No one ever thinks that negative values would occur for these variables. For example, we would not consider the example game being played for a minus number of trials, nor do we think of a negative number of shipments, negative pupils in a classroom, negative vehicles on a street, or minus prices on a stock exchange. Unless a computer is specifically instructed never to display a negative value for a variable, it will create negative values whenever the inputs indicate, no matter how wrong the inputs mights be. A frequent instruction in simulation programs is a check for and correction of negative or zero values that are otherwise meaningless as representations of object systems. Special care must be taken with some computers because in them zero is a positive number.

UNKNOWN INPUT

A user should always be confident about the data on which output is based. Unless an output record of the input is made by the computer itself, the user may wonder exactly what data the model was using. Since cards can become intermixed and computers can make reading errors, an output record of what the computer "thinks" the input data is can save time in explaining or debugging unusual results.

TOO MANY RUNS

The possible combinations of input, operations, controllables, and uncontrollables that might be tested by means of simulation is sometimes impractically large. Unless steps are taken to reduce the number of alternative simulation runs, a great amount of expensive computer time may be used up unproductively. Another problem is generating so much data that reaching conclusions becomes very difficult. To reduce the number of alternative simulation runs to be explored, simulation users may rely on judgment or on systematic search procedures that avoid going in ineffective directions. However, wherever an alternative remains unexplored, there is always the possibility it may have been important unless sound analytical reasons for ignoring it are available.

REDUNDANT COMPUTING

Professional organizations seldom allow model builders to do programming. The reason is that a model builder's concern with an object system may lead him to overlook computing efficiency. Professional programmers, on the other hand, are concerned with efficient operation of the computer, and usually do not care about effective representation of an object system. One example of these conflicting approaches is redundant computing. Redundant computing occurs when the model builder expresses a set of computations that generates identical values at each cycle because that is the way he sees what happens in the object system. In this case, the values could have been generated once at the beginning of the run, thereby saving the redundant computing in each cycle. The moral here is never compute anything within a "loop" (i.e., within a cycle or iteration) that can be computed outside of it.

REDUNDANT AUXILIARY ACCESSING

Input and output require a great deal more computer time than computation. If the operating system, the program, and the data do not use up all of the primary storage available to the programmer, then the most frequently accessed data and program segments in auxiliary storage should be transferred to primary storage. This technique can bring about extraordinary reductions in computer time. Sometimes additional run-time savings can be made by incorporating otherwise separate segments into the main program segment thereby avoiding the systems overhead used up to "call" the separate segments.

NO TEST CRITERIA

Unless some standard is set for performance of the computer program, the user may waste debugging time. The major requirement in debugging is a desk-tested example that will indicate whether the program is running properly. This example should be complex enough to use all the features step by step. Simple checks for extreme or unusual values can be made quickly. A "summary" prior-tested variable whose value depends on almost all input and operations is perhaps the fastest indicator whether a program ran as desired.

INVOLUNTARY STOPPING

Programs frequently run away with themselves and use up expensive computer time. Computer operators have learned to read the patterns of flashing lights to tell whether a program has fallen into a set of repetitive instructions from which it cannot escape without operator intervention. Simple counters and tests for maximum values can prevent losing a run for these reasons. Thus when trouble occurs, the program may be transferred to other segments that will run properly giving the user partial results. Otherwise, all is usually lost when the computer's operator or system switches to the next user's program.

Another way runs are lost is by insufficient or incorrect data. When the computer attempts to read cards or tape records that are either not there or are in the wrong form, most systems simply skip to the next program. Careful input data checking is the only way to avoid this problem. Even if options are available when input problems occur, they may lead to segments that require the data that did not arrive with the result of a lost run or bizarre output.

INCORRECT ACCESSING

Accessing the wrong variable can cause surprising results. When the address or the index number of a variable is computed within a program, an error in this computation can cause subsequent steps in the program to retrieve an unplanned value from data storage. This value may cause fatal difficulties. Tests to keep computed indexes, subscripts, and addresses within proper ranges can be built into programs, but few programmers do this.

INCORRECT STORING

Accessing data storage to obtain a value does not destroy the value in data storage. However, storing a new value destroys the old value in that particular storage location. When incorrectly computed indexes, subscripts, or addresses cause the program to write correct values in improper places, anything might happen. This problem became so great that the computer industry developed special hardware devices to prevent overlaying the nucleus of operating systems. Without such protection, any user might accidentally destroy the operating system, in which case all operations must start over from a system bootstrap.

PROGRAMMING TOO SPECIFIC

A computer model may be written for a specific single simulation run. Then to run with changed operations or with changed data may require extensive reprogramming. Reprogramming can be minimized by giving original computer programs as much flexibility as possible. This means most constants or controls are read as input and are not built into the program itself. It also means program segments are independently organized and clearly identified so that operations or segments to be changed may be easily found.

INCORRECT SYSTEM CARDS

At most installations, specific systems cards must be prepared by the user in addition to his program and data. From remote terminals, specific time-sharing procedures must be followed to load a user's program and data. A frequent cause of lost all-computer runs is incorrect use of these systems controls. When a user must state a maximum amount of computer time and a maximum amount of printed output, exceeding these limits terminates the program. Thus, for the sake of a few seconds more computing, or a few more lines of printing, a lengthy run may be lost and the computer time already correctly used must be duplicated. This means estimates of real run time are even more important for all-computer than for man-model simulation. In man-model simulations human adaptability can salvage a run while in all-computer simulations a violation of the inexorable rules of the operating system can effectively destroy a run and waste a turnaround.

Documentation

We have previously discussed the necessity for documenting man-model and man-computer simulations so that persons other than the original designer can use them. This is particularly important for computer programs. The objectives of computer program documentation are to provide:

1. A written statement of what the program directs the computer to do and how the computer is to go about this.
2. A written record of what the computer will expect as input and what it will produce as output.
3. A means of communicating with other persons who may run the program. Documentation is intended for users and operators, not for consumption by computers.
4. Specific run-time instructions for the computer operator, for the simulation user (who may be located at a remote terminal), and for recovery from any unusual situations at either of these locations.

Documentation of computer programs is perhaps one of the most economically important aspects of the computing industry, yet it is the one that is the least creative and sometimes the most neglected. Without documentation, valuable programs may in effect disappear on the death or resignation of an employee because no one else knows how to run them. Documentation is sufficient only when strangers to the project can take over without having to recreate all or portions of the project for themselves.

EXERCISES

7.1. Construct a flow diagram of a historical simulation of one of the man-component simulations you designed for past chapters.

7.2. Considering the historical simulation of Exercise 7.1, construct flow diagrams of the three additional steps for a complete simulation study listed on page 114. Include the specific input and model changes you would make.

7.3. Using the entire simulation from an exercise in a past chapter as the operations in a single rectangle, show by a flow diagram how you would go about management by simulation.

7.4. Write a short essay contrasting your work for Exercise 7.3 with online real-time process control.

7.5. Write a short essay on the distinction between stochastic simulation and a probabilistic analytical model.

7.6. Develop a flow diagram for a simulation model that is sequentially dependent on only a portion of the generated variables. Arrange for the unneeded data to be stored in auxiliary storage.

7.7. Add debugging variables to a flow diagram of one of your past simulation models. Would you remove them all when debugging was accomplished? What constitutes a state history?

7.8. A popular speaker on the subject of computers and simulation states that one of the goals of model builders is to achieve maximum parsimony with minimum distortion. These desiderata are in conflict: the more brief the model, the more it might misrepresent the object system; the truer the representation, the more detail required in the model. Write a short essay on (or be prepared to discuss) how a model builder might select criteria for balancing his desires for brevity in his model against his desires for accuracy and validity in his representation of the object system. Should a model builder change his purposes for the sake of parsimony?

7.9. Consider that the object system is some process in the real world, and that the model is an expression of the theory of this process. In this situation, we have a model of a theory of a process. Write a short essay on (or be prepared to discuss) whether a simulation model during execution is a model of a process or is actually the object system process itself. How does the human mind ever know anything at all about a process?

7.10. A test of the proposition that computers can think was developed by a famous mathematician named Turing. This test is now known as Turing's test. In this test, a human interrogator asks questions of a human respondent and of a computer. He receives written replies and can tell respondent A's answers from respondent B's answers, but he does not know which is the computer. If the human interrogator cannot tell whether A or B is the computer, then it might be said that computers can think.

Consider having written an all-computer simulation model of one of the participants in one of your prior man-component simulations, or of one of the bidders in the three-person automobile repair bidding object system of Exercise 4.2. Turing's test would determine whether the output of your all-computer simulation of this participant is indistinguishable from his real output. Write a short essay on (or be prepared to discuss whether passing of Turing's test is sufficient to validate your all-computer simulation model of this live participant. Should the operations of the model itself be considered? Is Turing's test appropriate if this participant's behavior is sequentially dependent on the behavior of other participants? Are there any alternatives to Turing's test for validating all-computer simulations of human behavior? Should additional live behavior of this participant, or even the behavior of other participants, be used to validate your all-computer model?

8

MONTE CARLO TECHNIQUES

Stochastic simulation was defined in Chapter 7 as the execution of simulation models whose output may vary trial to trial without trial-to-trial changes in input. Monte Carlo techniques are means for accomplishing stochastic simulation. Monte Carlo techniques may be used in all-computer simulations or in the computer component of man-computer simulations. They have uses other than in simulation.

Through the ages man has used chance processes. Theories of chance processes were known in ancient times. The evidence is the casting of the astragal bones from the ankles of sheep and goats in the manner of dice to subdivide land holdings. This was called casting lots. However, it was not until the 17th century that probability theory became formalized. Interestingly, this mathematical development arose out of inquiries into games of chance—a fact that makes our use of a penny-matching game for illustrative purposes rather appropriate.

Another important development occurred in the 1940's during war work on the atomic bomb. Scientists were frustrated by being unable to solve some complex nonprobabilistic mathematical problems directly by analysis. By combining chance processes and probability theory, they solved the problems indirectly. To do this, they created a new problem stated in terms of stochastic models in such a way that the new problem was equivalent to the complex nonprobabilistic problem. If the first problem was directly unsolvable, the stochastic version was even more so. But the stochastic version represented chance processes that could actually be carried out. This is what the scientists did. They actually performed the chance processes represented by the stochastic formulas in the new version of their problem. Their methods were equivalent to flipping coins or throwing dice, but much more efficient. The results of these actual experiments were then substituted for the stochastic expressions of the new version of the problem to give answers to the original nonprobabilistic problem. These scientists gave the name *Monte Carlo* to their new technique.

Monte Carlo Estimates in the Unit Square

Monte Carlo techniques were born when probability theory and actual chance processes were used to solve problems that had no stochastic aspects at all. This situation is easily illustrated by the problem of estimating the area of an irregular surface such as a lake. Consider the following diagrams:

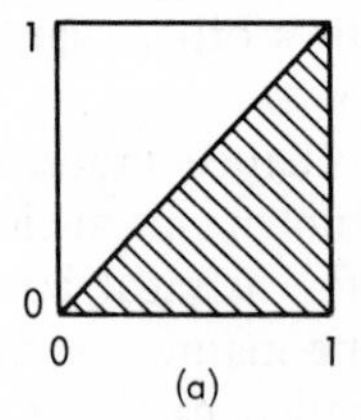

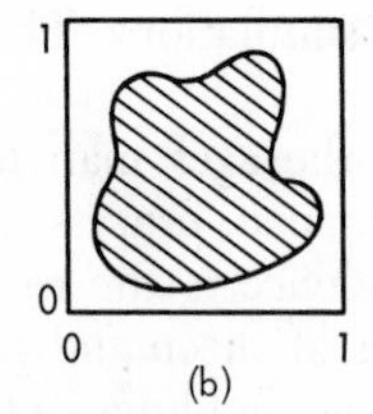

FIGURE 8.1. Unit Squares

The rectangles in Figure 8.1 are called unit squares. They are drawn so that the length of each side is one. Hence any point along that length can be located by a number between zero and one. Assume we have two such numbers. Let the first locate a point on the bottom of the square, and the second locate a point on the left side of the square. Then any point *within* the square can be located as being vertically above the point on the bottom and horizontally to the right of the point on the left side. Since there are an infinite number of decimal fractions between zero and one (more precisely, between 0.0000 . . . and 0.9999 . . .), we can locate an infinite number of points along the bottom and an infinite number along the left side. This provides an infinity of points within the unit square. An infinity of points means we can find as many different ones as we wish.

The area of the unit square is exactly one square unit (which might be thought of as a square inch, square foot, square mile, etc.). The shaded part of Figure 8.1(a) is divided from the balance of the unit square by a diagonal. Its area is exactly one-half, which we know both from geometry and from mathematical analysis. However, the shaded part in Figure 8.1(b) is irregular, and we have no geometry or mathematical analysis to compute its area.

Given any point in the unit square, let us assume we have formulas which would tell us whether the point is inside either of the shaded parts of Figure 8.1. Formulas are not absolutely necessary for what we intend to do, but they make our decisions about the locations of points much more rigorous.

The Monte Carlo procedure to find the area of a shaded part is simply the selection of a large number of random points within the unit square. This is done by repeatedly choosing two random numbers between zero and one, using the first to identify a point on the bottom and the second to identify a point on the left side. These two points identify a single point within the unit square. This point is then tested, either by formula or by inspection, to see if it is inside or outside the shaded part and the result recorded. After a predetermined number of points have been tested, we compute the proportion of points falling inside the shaded part and use this proportion as an estimate of the area of the shaded part.

We could carry out the above Monte Carlo procedure by drawing chips from a hat. Assume there are ten chips of identical size in a hat with the numbers 0, 1, 2, . . ., 9 marked on them. We could shake up the chips, draw one, note its number, replace it in the hat, reshake the hat, and draw another. By dividing by ten, we could use the first one to locate a point on the bottom of the square, the second to locate a point up the left side of the square. Then using these two points, we could locate a point within the square and test it to see if it fell inside or outside the shaded part. We could then repeat drawing pairs of digits from the hat until we had located the predetermined number of points in the unit square from which the area is to be estimated.

Someone might object to the above procedure because only ten possible points can be located on the bottom and on the left side. Moreover, they are biased to the left because no points beyond .9 would ever be identified. To correct this condition we might place one hundred chips in the hat and number them 0, 1, 2, . . ., 99 to obtain more points in a unit length (dividing by 100). However, no point greater than .99 and less than 1.00 would ever be chosen. For more precision, we might place 1,000 chips in the hat numbered 0, 1, 2, . . ., 999, thereby providing 1,000 possible points that might by chosen within a unit length (dividing by 1,000). Still, there remains a slight bias to the left because no point greater than .999 and less than 1.000 would ever be identified. The number of chips in the hat might be increased, but this requires a rather large hat and the procedure of shaking it up becomes physically burdensome.

Random number generators that do not require hats were mentioned in the last chapter. Let us now assume that we have a random number generator that will provide numbers between zero and one in a way that, if we chose an infinite number of them (which we could never do because infinity means we would never stop choosing numbers), the proportion of those numbers falling between any subinterval in the unit length would exactly equal the proportion falling within any other subinterval of equal size. When this and other characteristics to be discussed later are met, the numbers are called uniformly distributed random numbers between zero and one. Let us further assume that this random number generator is a handy part of a computer program, but not, you will recall, a part of the computer model of an object system. Let us also imagine we have programmed as computer models the two unit squares of Figure 8.1. These models are merely the formulas by which the decision is made whether any point falls inside or outside the shaded part.

The computer is now at our service to select pairs of random numbers very rapidly, each pair identifying a single random point within the unit square. We could select one hundred points and count how many of them fell within the shaded part. We might claim this number, divided by one hundred, to be an estimate of the area of the shaded part.

However, we might feel uneasy about only one hundred points because, purely by chance, we may have selected combinations of numbers that do not appear uniformly distributed. Figure 8.1(a) gives an estimate of this bias. If we chose many points, we would expect close to half of them to fall within the shaded part. But with only 100 points, it is possible that as few as ten, or fifteen, or twenty might fall in the shaded part. On the other hand, as we tested thousands and thousands of random points, we would expect the proportion falling within the shaded part of Figure 8.1(a) to approach more closely one-half. But we must stop testing random points sometime. This may be done when large blocks of random points no longer change our estimate of the area very much. When we finally stop, we take the estimate of the area of the shaded part to be the number of points falling within it divided by the total number of points generated. This is a proportion between zero and one. The area of the unit square is exactly one. Hence, the number that represents the proportion also represents the area of the shaded part. If the square had an original area other than one, we would merely multiply that area by the proportion to obtain the desired estimate.

While for Figure 8.1(a) we expect our Monte Carlo estimate to be very close to one-half, we have no similar expectation for Figure 8.1(b) because mathematical analysis fails us. All we have is the proportion generated by Monte Carlo techniques. It is important to note that this was a great deal of work.

For Figure 8.1(a), we could have computed the area of the shaded part directly by analysis. The considerable extra effort required to make a Monte Carlo estimate was not needed (except to illustrate the technique). However, for Figure 8.1(b), the Monte Carlo estimate is all we have. Here, we see the economic source of the moral: never simulate when analysis will suffice.

Monte Carlo Simulation of Coin Tossing

Let us apply the above ideas to the chance processes of the example game. This is, of course, only for the versions where one or both players actually flip a penny. To simulate the mental processes of subjects who make decisions by conscious deliberations, Monte

Carlo techniques would be used only where subjects demonstrated some form of random behavior.

To simulate coin tossing we require a means for determining at each trial which face of the coin appears. A "fair" coin is one we believe has equal chances of either a head or tail appearing if tossed "fairly." To represent tossing a fair coin, we could use the same Monte Carlo technique used to estimate the area of the shaded part of Figure 8.1(a). For each simulated toss of a coin, we would simply generate two random numbers between zero and one and translate these into a random point within the unit square. If the point fell within the shaded part, we would declare that a tail appeared. If the point fell in the unshaded area, we would declare that a head appeared. Here, we are using the same scheme scientists created to estimate nonprobabilistic values such as areas to simulate chance processes for which a stochastic model is appropriate.

The example game involves tossing two coins and testing whether they match or not. If we believe by analysis that there is a 50-50 chance of their matching, we could still use the Monte Carlo technique we applied to Figure 8.1(a). We would merely substitute the labels "match" or "no match" for "head" and "tail." If we are not confident of equal chances for "match" or "no match," things become more complex. In this case, we would want to represent the *compound* chance process of flipping two coins in a way that reflects all possible ways the two outcomes may occur and associate with each way a probability measure. While there are only two ultimate outcomes, match or not, these can occur in four different ways. The flipping of each coin is itself a separate chance process. Since this process cannot further be broken down, we call it an *elementary chance process*. Elementary processes generate elementary events. Flipping both coins together, as in the example game, creates a *compound chance process* with results we call *compound events, compound outcomes*, or simply *outcomes*.

Sometimes compound processes can be directly analyzed in terms of their elementary processes. For example, for each way Player R's coin may show, Player C's coin may show in two ways. The compound events are: (H,H); (H,T); (T,H); (T,T). Figure 8.2(a) shows how we might represent these four outcomes in the unit square. (We now discover why we have been using the names Player R and Player C. In this grid, Player R controls the choice of a *r*ow and Player C controls the choice of a *c*olumn.) Figure 8.2(b) represents the example game when Player C throws the bent penny

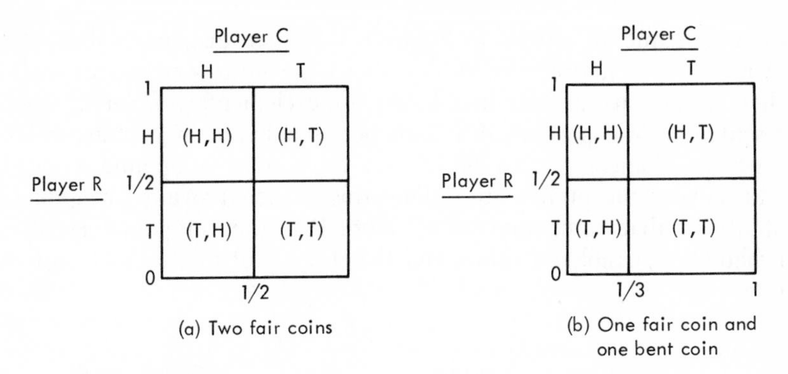

FIGURE 8.2. Area Representations of Tossing Two Coins

with probability of tails equal to two-thirds. Probability of tails equal to two-thirds is represented by assigning two-thirds of the area under Player C's control to tails.

When we generate the first random number and locate a point on the bottom, we have simulated Player C's coin; when we generate the second random number and locate a point on the left side, we have simulated Player R's coin. Representing these compound events by areas is inefficient because two random numbers must be generated to determine a single outcome. We shall show below how compound events can be simulated by a single random number.

Tree Diagram Representations of Example Game

Figure 8.2 represents graphically the multiplication rule of probability theory. This rule states that the probability of a compound event may be found by multiplying together the probabilities of the separate independent elementary events that make up the compound event. The resulting probabilities still add up to one, just as the separate partitions in Figure 8.2 exhaust the areas of the unit squares. This means at least one of the compound events happens at each trial.

Events are independent when the process generating them is not influenced by events generated in the past. The multiplication rule of probability theory for independent events can also be shown by a tree diagram. Probability tree diagrams for single trials of the example game are shown in Figure 8.3. In tree diagrams, labels of elementary events may be shown above the limbs of the tree and their elementary probabilities below. One elementary event is represented by the portion of a limb between branching points. Following a branching path to its end traces one compound event. Multiplying the probabilities along this branch provides the probability of that compound event. Note that the compound events exhaust all possible outcomes and that their probabilities add up to one.

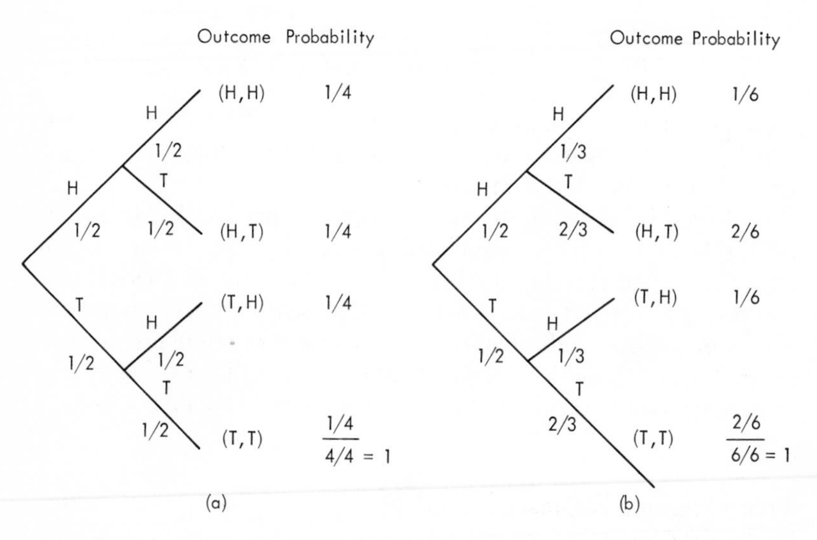

FIGURE 8.3. Probability Tree Diagram Representations of Tossing Two Coins

Matching in the example game is an outcome that occurs in two ways: (H,H) and (T,T). These are two compound events. Under the addition rule of probability theory, we may add the probabilities of all mutually exclusive ways in which an outcome may occur to obtain its total probability. These ways may be elementary events, compound events, or both, depending on the rules for defining outcomes. In the example game, by the addition rules, the probability of a match is the sum of the probabilities for (H,H) and (T,T). Similar analysis provides the probability for not matching. Since "match" or "no match" exhausts the outcomes, and they are mutually exclusive, the probabilities of the two outcomes add to one. This is precisely the analytical model of the version of the example game in which both coins are actually tossed. The tree diagrams of Figure 8.3 are representations of this analytical model.

Cumulative Probabilities

Probabilities are numbers that add up to one. If we go down a list of mutually exclusive and exhaustive outcomes and accumulate the probabilities associated with each, when we reach the bottom of the list, our accumulated probabilities will total one. This has been done in Figures 8.4(a) and (b). In these tables, the outcomes have been rearranged slightly so that the two compound events that represent "match" are shown together, and the two that represent "no match" are shown together. Columns (1) in Figures 8.4(a) and (b) show the probabilities of the compound events from Figure 8.3. Columns (2) show the cumulative probabilities as these probabilities are summed over the four compound events. Columns (3) show the cumulative probabilities when only the two outcomes "match" or "no match" are considered. It is important to notice that the individual probabilities of columns (1) sum to one, but that the cumulative probabilities of columns (2) and (3) terminate with one.

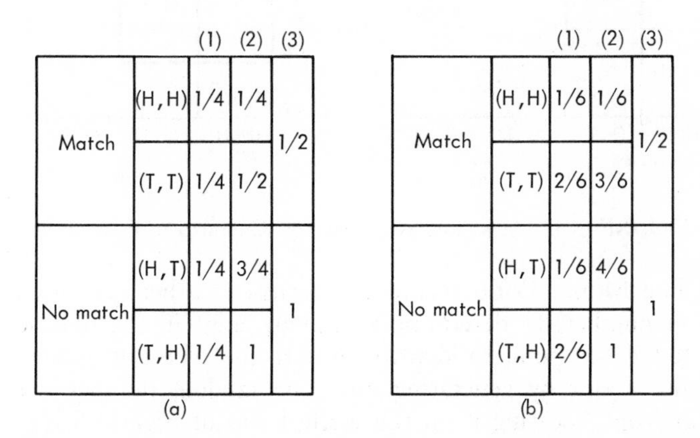

FIGURE 8.4. Cumulative Probabilities of Tossing Two Coins

Recalling that the unit square has area equal to one, we can establish an analogy of cumulative probabilities with the unit square. This has been done in Figure 8.5 in which the separate areas representing the probabilities of compound events from Figure 8.2 have been added graphically to accumulate to a total area of one. Area in Figure 8.5 is measured on the vertical axis, which is a unit length.

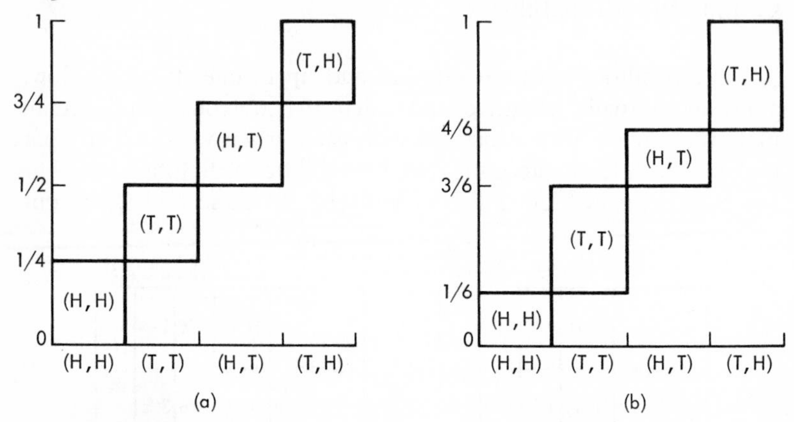

FIGURE 8.5. Area Representation of Cumulative Probabilities

The Monte Carlo technique discussed earlier generated two random numbers to determine a random point in the unit square, then tested this point to identify an outcome. We can now achieve the same result by generating only one random number between zero and one, locating it on the vertical axis of Figures 8.5(a) and (b), and translating this single location on a unit length into an outcome. The translation is done by looking horizontally to the right of the point and finding the corresponding area of the unit square (shown cumulatively to the right) which identifies a particular compound event. To gain efficiency, we have graphically combined analysis and simulation.

Generalized Random Outcome Generation

Generating outcomes with only one random number at each trial can also be done numerically. This is shown graphically in Figure 8.6 where segments of a unit length are identified in terms of the possible compound events, which are in turn translated into outcomes that may occur on a single trial of the example game.

Conceptually we can now see how simply the Monte Carlo technique is. All we need to do is label segments of a unit length in

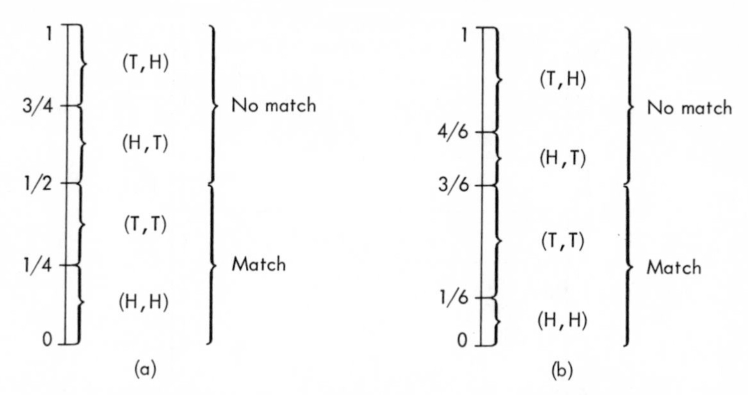

FIGURE 8.6. Identifying Events on the Unit Length

terms of outcomes. We must be careful to exhaust the entire unit length and also see that none of the labeled segments overlap. Then, we merely generate random points on that unit length and translate them into outcomes according to the segment they fall in. (Later, we shall also label individual points on the unit length so we can translate random points into points on a scale of outcomes.)

Discrete Probabilities

Through other books, the reader may investigate probability theory in any depth he wishes. Here, it is sufficient to understand only a few basic notions. One of these is the convention that probabilities are numbers that add up to one when associated with mutually exclusive events that exhaust all possibilities in a defined situation or trial. Under this convention, one of these events will be sure to happen at each trial or instance of the defined situation.

Probability of one represents certainty and means that one of the possible events will happen. Probability of zero represents impossibility. Thus, any possibility that may occur must have a probability number greater than zero. The convention that probabilities are numbers between zero and one is really a means for scaling uncertainty between impossibility and complete certainty. Under this convention, the requirement that the events be mutually exclusive means that when one occurs, another does not. If mutually exclusive events are not the case for a particular problem, events can always be made mutually exclusive by redefining them. For example, if two events partly overlap, such as events A and B in Figure

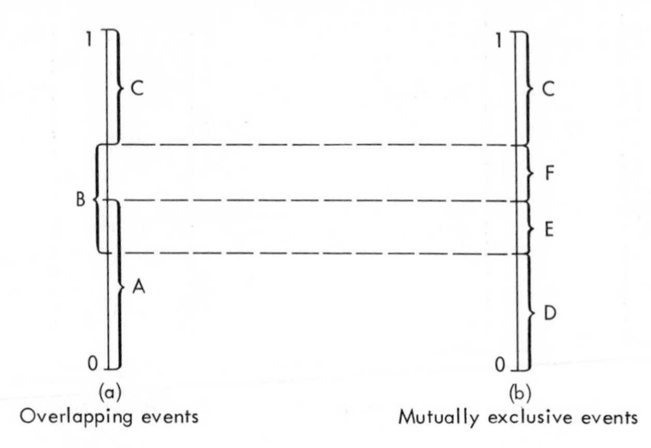

FIGURE 8.7. Exhaustive Events on the Unit Length

8.7(a), they can always be made mutually exclusive by redefining them, for example, into events D, E, and F in Figure 8.7(b). The redefined events will carry different probability numbers represented by different intervals on the unit length. If the redefined events exhaust all possibilities, these new probability numbers will total one and their representative intervals will fill up the unit length. In Figure 8.7(b), events C, D, E, and F are both mutually exclusive and exhaustive.

When the events themselves are distinct, as in Figure 8.7, a separate probability may be associated with each event. In this case, the probability numbers are called *discrete probabilities.* So far in this book we have dealt only with discrete probabilities. In the next section, we shall discuss probabilities that are appropriate when events cannot be identified as distinct intervals on the unit length.

Another way to represent discrete probabilities as areas of the unit square is shown in Figures 8.8(a) and (b). Figure 8.8(a) represents the probabilities of the number of heads appearing when two coins are tossed. Again, this is the compound chance process of our example game. Figure 8.8(a) is simply the piling of separate prior area representations of probability onto labels identifying the respective number of heads that area represents. Here, events are defined to be the number of heads appearing on the toss of two coins ($n = 2$). These are compound events because each coin is itself an elementary chance process.

If eight coins are tossed at each trial, the number of heads appearing may range between zero and eight. These are distinct com-

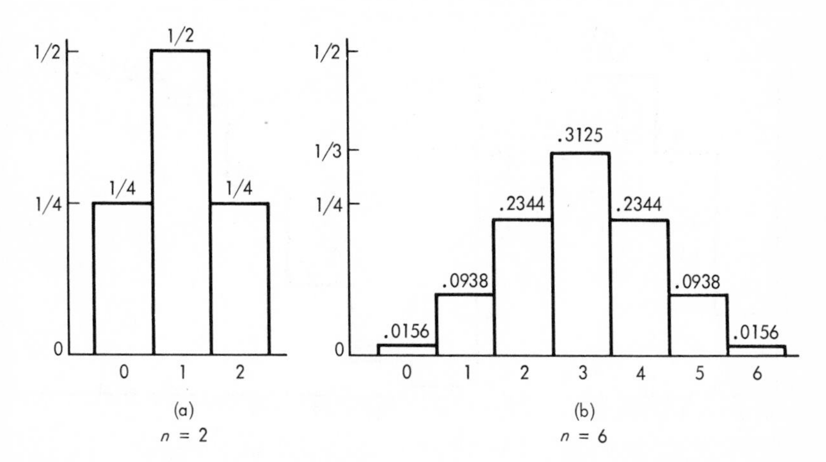

FIGURE 8.8. Discrete Probabilities of Numbers of Heads Appearing when n Coins are Tossed

pound events requiring discrete probabilities. Figure 8.8(b) represents the distribution of discrete probabilities over the nine possibilities.

The total area in each of the two diagrams of Figure 8.8 represents one, and thus these areas would fill a unit square. If we accumulate these rectangles, they will stack up to a height of one. If we offset each rectangle as we stack it, as we did in Figures 8.5, and then throw away the bottom and the right side, we can display cumulative discrete probabilities as shown in Figure 8.9. The zigzag graph of cumulative discrete probabilities in Figure 8.9 easily identifies points on the vertical unit length with events on the horizontal axis. With these graphs, any generated random number between zero and one can identify one of the mutually exclusive and exhaustive distinct events merely by locating the point on the vertical unit length, looking horizontally to the right for the left side of a rectangle, and finding the label of the event below the top of that rectangle. Since at least one of the events must occur, we shall let zero, if it appears as a generated random number, represent the first event.

Continuous Probabilities

So far events have been identified in terms of distinct categories each represented by a discrete interval on the unit length.

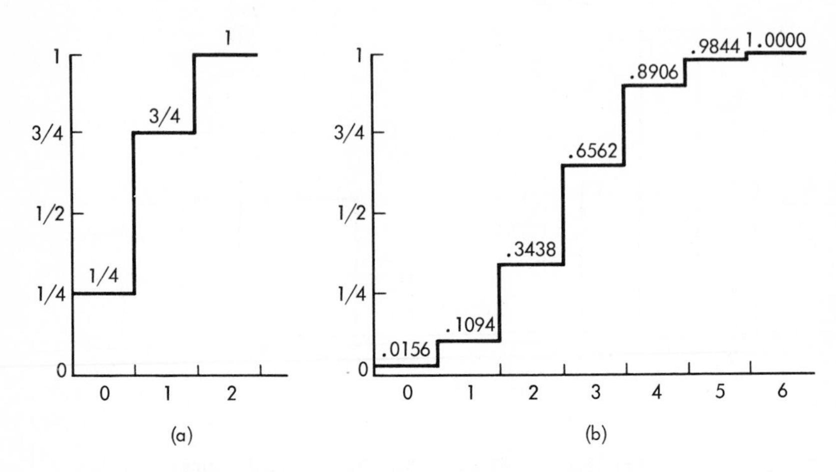

FIGURE 8.9. Cumulative Discrete Probabilities

Events may also be stated in terms of length, time, size, weight, degrees, or in similar dimensions that are numbers from a continuum of numbers. In any interval of a continuum, there is an infinity of numbers, and hence an infinity of possible events. If we assign some positive probability number to each of the infinitely many possible events, our probabilities would add up to infinity. This violates the convention that probability numbers add up to one. Suddenly, serious conceptual problems arise when distinct identification of events is not appropriate.

One thing we could do to overcome this difficulty would be to use a discrete probability scheme with thousands or millions of distinct events identified on a unit length. In this case, the diagrams (b) in Figures 8.8 and 8.9 (which are for only eight events) would approach smooth curves. In fact, as the number of coins tossed approaches infinity, the shape of curves such as that of 8.8(b) approaches a well-known family of curves called *normal curves.* For such a large number of tosses, the curve of 8.8(b) would look very low and very wide.

Normal curves may take many locations and shapes. One of this family of curves has been standardized and documented. It is called the standard normal curve and is shown in Figure 8.10(a). It can be thought of as errors of measurement, some being on the high side (above zero) and some on the low side (below zero). Details of the standard normal curve can be found in books on statis-

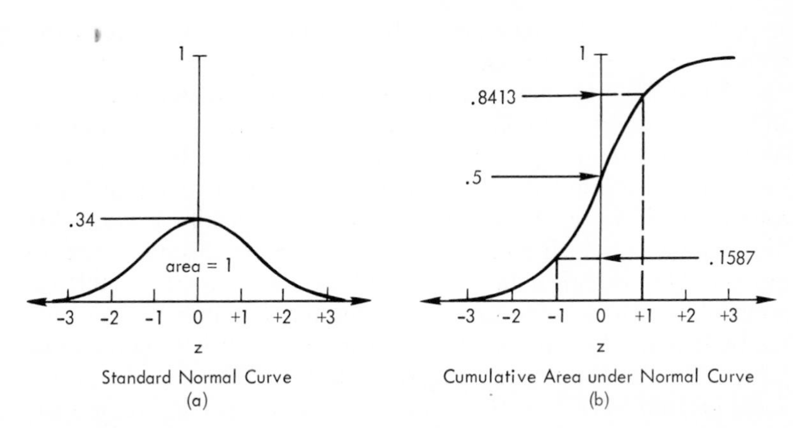

FIGURE 8.10. Continuous Probabilities of the Standard Normal Curve

tics and probability. For our purposes, we should particularly note that the horizontal axis is a continuum from minus infinity to plus infinity. Points on this axis are called z-values or z-events. Each point is a different event and not one of a category of points identifying a single distinct event. We should also note that the standard normal curve has an area under it equal to one, and that half of that area is above zero and half below.

At first glance, it would seem that we could read the probability of any z-event from a diagram of the normal curve just as we can read the probability of distinct events from area representations of discrete probabilities. Unfortunately, this is not the case because any area representation under the normal curve represents an infinity of z-events on the horizontal continuum. To reduce the area to a single z-event, which is a point on a line, the area above it is reduced to zero, which when interpreted as a probability means the z-event is impossible. If we were to identify all of the infinitely possible z-events and add up their zero probabilities, we would obtain zero and should conclude that nothing ever happens. On the other hand, if we associate with each z-event a positive probability number, these probability numbers would add up to infinitely in violation of the convention that they must add up to one.

This dilemma is solved by identifying areas under the normal curve with subsets of z-events. These subsets of z-events are identified as all the z-events falling between two z-values on the horizontal axis. For example, using tables of the normal curve, the probability of a z-event falling between z-value of minus one and z-value of plus one is .6826. This probability is simply the area under the normal curve between these two z-values.

Since the area under the normal curve is one, and since probabilities must add up to one, we can accumulate the area under such curves from left to right. If we do this, we start with zero area at the left and accumulate as we move to the right along the horizontal axis. If we go far enough, we will have finally accumulated the entire area, which equals one. We can graph this accumulative process. This is shown in Figure 8.10(b). With these cumulative probabilities which begin in zero and terminate in one, we can state the probability that any event on a continuum will fall between two values. This is done by finding the difference between the cumulative area up to each of these two values. For example, we find the probability that a z-event falls between the z-values -1 and $+1$ by subtracting from the cumulative area up to $z = +1$ the cumulative area up to $z = -1$.

For both discrete and continuous probabilities, the vertical unit length is itself a continuum. For distinct events, the horizontal axis is simply a system for labeling categories of points. For continuous events, the horizontal axis is itself a continuum that labels point-events. In either case, a simulated event can be generated by first locating a cumulative probability point on the vertical unit length. Then, for distinct events we translate this point into an event label by means of discrete intervals or the zig-zag graph. For continuous events, we translate the vertical axis point into an event on the horizontal event continuum by means of the curve. For example, a generated random number of .8413 identifies the z-event "+1." Similarly a generated random number of .1587 identifies the z-event "−1."

In effect we have turned probability theory onto itself. One of the goals of probability theory is to describe the distribution of probability numbers over events. Our goal to create events by generating probability numbers is just the reverse. While probability theory proceeds from events to probabilities, Monte Carlo simulation proceeds from probabilities to events. One purpose is the inverse of the other. Mathematically, formulas for proceeding from events to probabilities have been highly developed. Unfortunately, the mathematics of proceeding from probabilities to events is not as highly developed.

Of course, any shape curve may be drawn by hand and event-points on the horizontal axis can be identified by visual inspection.

This technique may be accurate enough for many purposes. However, it is time-consuming and cannot be computerized. For computer-based Monte Carlo simulation, it is necessary that the formula for the inverse function of going from probabilities to events be known in order to provide a means within the computer for identifying events from points on the unit length.

For discrete events with discrete probabilities, inverse formulas are not required. Identification of distinct events in computers is done by searching lists.

When inverse formulas for continuous probabilities are not available, a discrete probability approximation may be substituted. The disadvantage is that all event-points on the event continuum are not eligible for inclusion in the simulation. On the other hand, almost any continuous probability distribution can be approximated with discrete probabilities. This avoids restricting the simulation design to those mathematical formulations for which the inverse relationships are known. Another strategy is to use a substitute curve for which the inverse formula is known. This is done by obtaining the event from the substitute model, then translating that event back to the dimensions of the original model.

Flow Diagrams for Monte Carlo Simulation

Monte Carlo simulation is shown by flow diagrams in Figures 8.11 and 8.12. These diagrams represent Monte Carlo simulation either by hand or by computer. The logical steps for discrete simulation are the same in the computer as they are by hand. For continuous simulation, Figure 8.12, it is possible to use graphic representations for hand simulation, but inverse formulas are required for computer simulation.

Notice that in discrete simulation the test is whether the generated random number is "less than" the upper boundary number for each event on the unit length. Thus the random number zero, if generated, identifies the first event on the unit length, and the random number one, if generated, identifies the last event. (Usually, random number generators do not create a one, but the number .9999. . . .) This is a standard convention called "lower bound inclusive." We followed it earlier when we let zero be an event. Under it, the upper bound belongs to the next event as the search to identify a random number proceeds upward along the unit length.

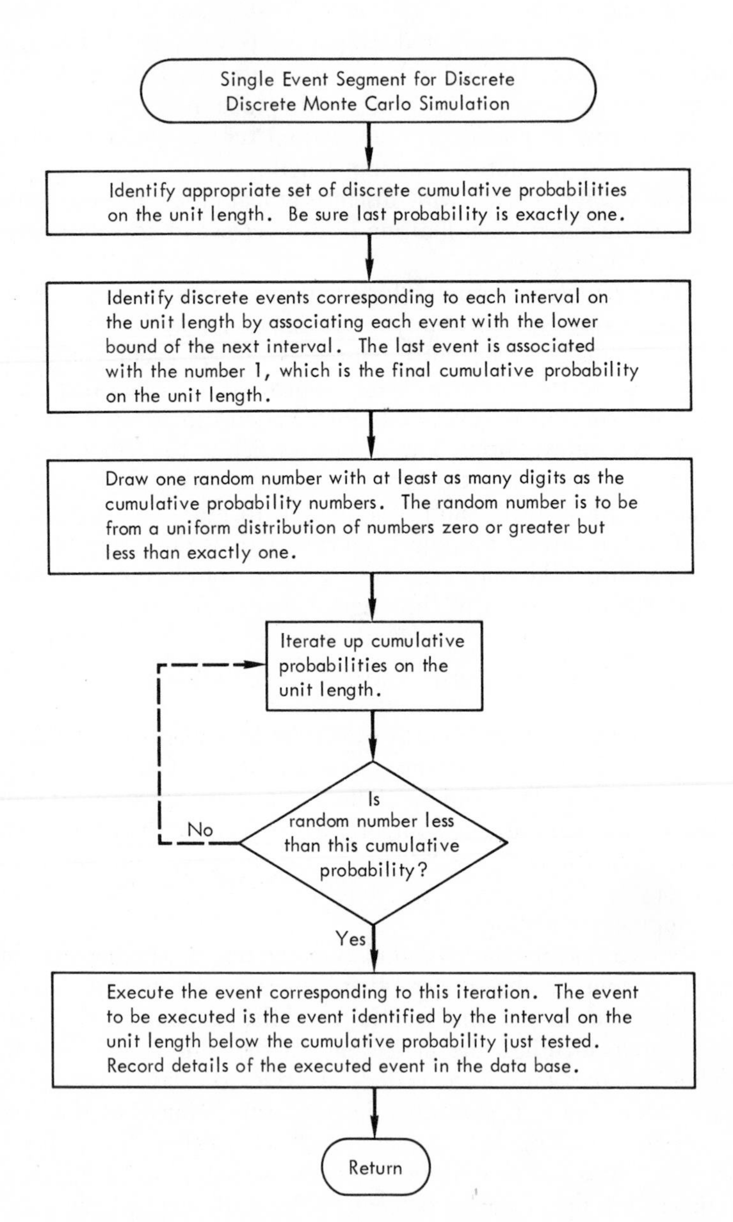

FIGURE 8.11. Single Event Discrete Monte Carlo Simulation

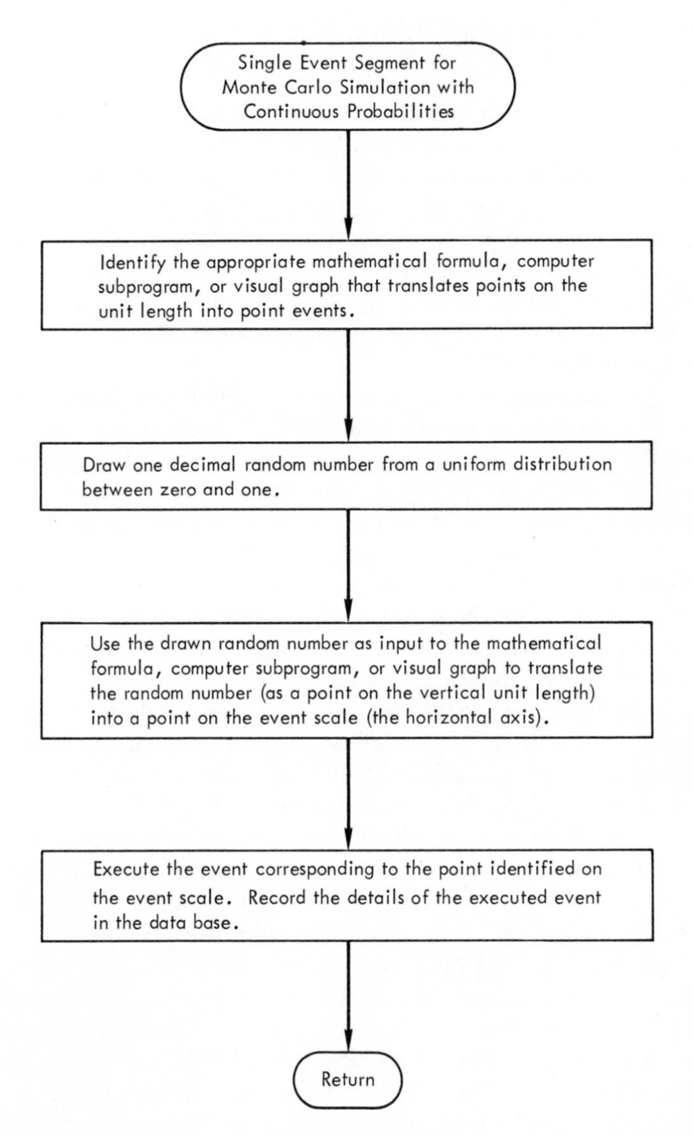

FIGURE 8.12. Single Event Generation by Monte Carlo Simulation with Continuous Probabilities

Sources of Probability Information

Usually, probabilities are developed from past experience with object systems. Sometimes, a combination of experience and reason leads to generalizations that a probability distribution ought to look a certain way. Eventually some of these generalizations become overpowering assumptions always present in our thinking about a particular chance process. One such overpowering assumption is "the chances are so small" argument by which many persons rationalize venturing onto highways where thousands of persons are killed each year.

Another overpowering assumption is that the probability of heads on the toss of a coin is one-half. Despite this overpowering assumption, we would be very surprised if actual observation of a coin tossing experiment resulted in heads exactly half the time. Nevertheless, we continue to use the assumption as a basis for decisions, even without testing coins for fairness. Do football officials toss-test their coins before games? More importantly, how would they decide whether the coin passed the test? Such overpowering assumptions are a part of our culture. They are also important sources of probability information.

Theorists in statistics and probability have developed several discrete and continuous probability distributions that are now taken as standard representations of chance processes. The more important of these and their applications are shown in Figure 8.13.

When historical observations do not seem to fit into a standard mold, Monte Carlo techniques offer a direct answer. The historical distribution itself can be used to represent the chance process. By dividing the observations into subclasses, counting the cases in each subclass, and dividing by the number of observations, the relative occurrence in each subclass is computed. These relative frequencies may then be taken as the probabilities of the events. If the subclasses are distinct events, the probabilities form discrete probability distributions that can be accumulated along a unit length. If the subclasses are intervals on a continuum, then some one point in the interval must be selected as the event. Usually, the midpoint of the interval is taken to be the event. This procedure is an approximation of continuous probabilities by using discrete probabilities. The approximation can be made as fine as desired by

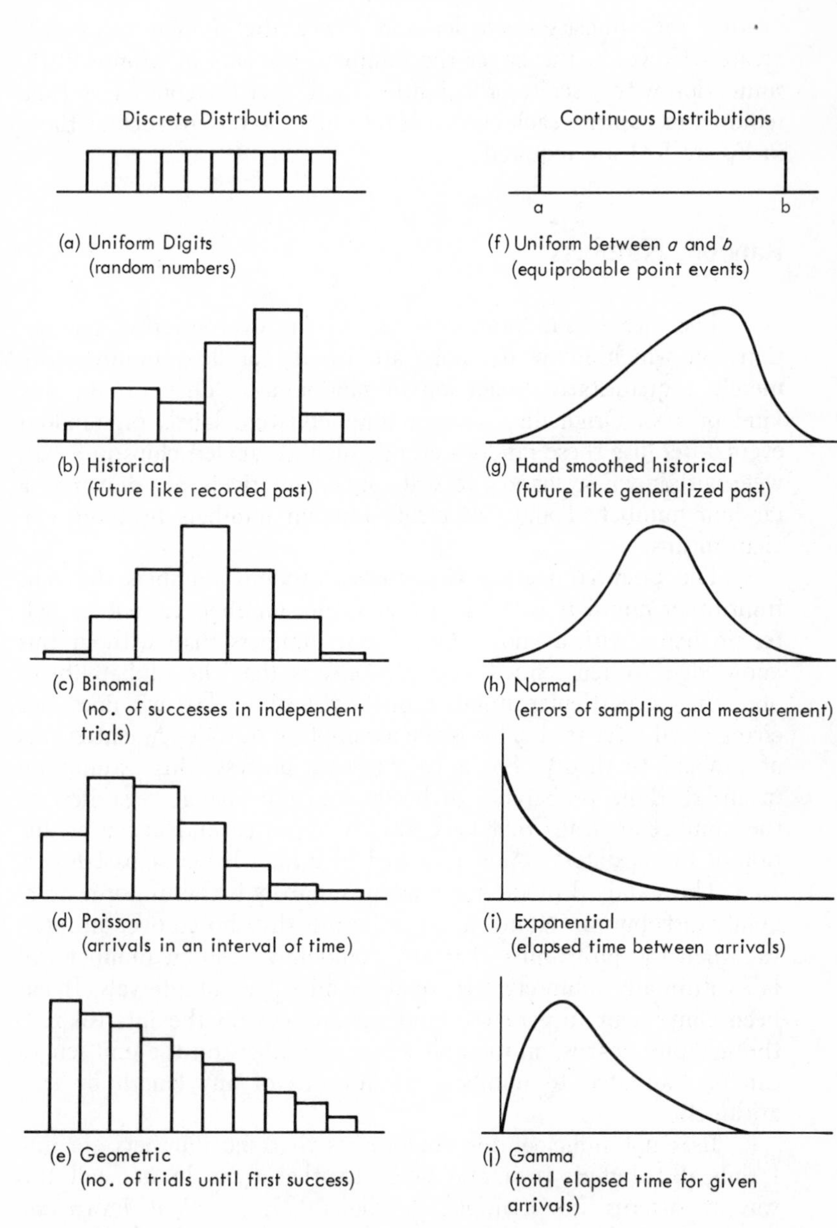

FIGURE 8.13. Example of Probability Distributions (with Illustrative Application)

making the subclasses smaller and hence the number of events greater. However, the larger the number of events in Monte Carlo simulation with discrete probabilities, the longer the computing time required to identify each event because more iterations such as those in Figure 8.11 are required.

Random Numbers

The idea of randomness is one of the overpowering assumptions on which many decisions are based. Random numbers are merely a quantitative reflection of randomness. They can be any kind or size. Originally, random numbers were labels on random events. Because these random events, such as labeled chips in a hat, were withdrawn by hand, we still speak figuratively of "drawing" a random number. Today, we create random numbers by more efficient means.

The essential feature that makes random numbers different from other numbers is the idea that future numbers cannot be better predicted with a knowledge of past numbers than without this knowledge. When refined, this idea asserts that the probability of obtaining a particular number on any trial or "draw" does not change trial after trial. This is the assumption of independent events of probability theory. For a coin tossing process, this assumption means that the probability of heads does not change regardless of the number of heads that have already appeared and hence prediction of the next toss is not improved by a knowledge of past tosses.

The standard model for random numbers is the uniform probability distribution. Numbers are uniformly distributed over an interval when the probability that any "randomly" chosen number will fall within any subinterval is equal for all equal subintervals. It has been convenient to generate random numbers on the interval zero through one, that is, on the unit length. Numbers on the unit length can be translated to numbers on intervals of any length by easy arithmetic.

It is not sufficient for randomness that the numbers are uniformly distributed. It is also necessary that they do not fall this way in patterns. For example, the digits 0, 1, . . ., 9, if drawn randomly, would be expected to appear about one-tenth of the time. However, they could appear one-tenth of the time by all zeros appearing first, all ones next, all twos, etc. If this were to happen, we would be able to predict exactly what number would appear next. This would violate the fundamental notion of randomness.

There are several standard tests of randomness. Some of these tests compare the characteristics of sets of individual decimal digits with the following expectations:

1. We expect random decimal digits to reappear at intervals that average ten digits in length (the *gap test*).
2. For small groups of random digits, we expect repetitions of patterns to occur with frequencies in accord with probability theory (the *poker test*).
3. We expect each combination of two successive random digits (00 to 99) to appear with equal frequency (the *serial test*).

Other tests compare sets of continuous decimal numbers (usually between zero and one) with the following expectations:

4. We expect no correlation to occur between successive numbers (the *correlation test*), or, equivalently, we expect random points in the unit square (each requiring two random numbers) to be evenly distributed.
5. We expect the average distance between two random points in the unit square to be one-half (the D^2 *test*).

Usually random numbers are generated in the computer by a mathematical formula which is repeatedly executed on its own output. Given a particular start, this generator will always output the same list of random numbers. If we knew this list, we could predict exactly which number would come next. The same is true for published lists of random numbers. If the person who makes the draw knows the list, the numbers are no longer random. Hence, randomness, for our purposes, is not only in the properties of the numbers drawn but also in the state of mind of the observer or "drawer."

Random numbers for which the exact list could be known are called *pseudo-random numbers*. Originally, random numbers were generated by actual chance processes rather than by mathematical generators. These processes produced physical and electronic outputs that were translated into numbers. For example, tossing of a coin is a real chance process that produces real random outputs having two values, either heads or tails, which can be interpreted as either zero or one. Conversely, a decimal pseudo-random number generator could be interpreted so that odd numbers represent heads or zeros and even numbers represent tails or ones. Many translations are possible. In fact, the entire Monte Carlo technique is a translation of uniform random numbers into other meanings.

0.54548335	0.43878794	0.72337675	0.39116907	0.83662391
0.49922121	0.46571195	0.30128038	0.61627400	0.98612070
0.37025809	0.34646153	0.74644649	0.36052489	0.44513071
0.42605996	0.55018330	0.46655953	0.84770727	0.88720751
0.69387960	0.17841017	0.82554424	0.34757400	0.65554559
0.80510724	0.93073332	0.33843434	0.65400565	0.87812412
0.38269413	0.39304709	0.91403508	0.94678664	0.45440423
0.20534551	0.14243436	0.00649624	0.75706804	0.48394215
0.09004080	0.18476474	0.29822123	0.12644494	0.07467782
0.31006253	0.18827474	0.33908570	0.34004152	0.98847806
0.87049425	0.32666230	0.12552536	0.81319129	0.74941897
0.17779207	0.32198119	0.33175802	0.09271681	0.57047844
0.58841896	0.39620805	0.08147699	0.92298937	0.80464327
0.52095520	0.88394105	0.61504948	0.73482764	0.87351990
0.62767053	0.90434361	0.77702618	0.52306473	0.14515269
0.16333282	0.67362285	0.57174146	0.36784291	0.06138418
0.05771821	0.79385161	0.24364603	0.31721127	0.71045339
0.40781903	0.05283337	0.64662874	0.40427232	0.60597491
0.99739885	0.53061843	0.20712042	0.46715593	0.93885207
0.42870843	0.12258160	0.87711370	0.15944791	0.06266403
0.94095242	0.08173788	0.02185562	0.39549232	0.17625380
0.49809122	0.40226316	0.93075824	0.96418023	0.40825725
0.77192140	0.95721316	0.79598641	0.16100025	0.80212331
0.36373699	0.96331191	0.50623846	0.36762297	0.64959157
0.58894217	0.68732846	0.82349062	0.75498736	0.11850828
0.91616333	0.43040526	0.33696187	0.14812326	0.85608268
0.80338597	0.11557180	0.46295691	0.73759520	0.25895894
0.91539633	0.16174686	0.73191428	0.93576384	0.02735382
0.74224818	0.20730495	0.56359589	0.51583052	0.02262001
0.49324453	0.75588727	0.09612250	0.77374959	0.77739513
0.70062411	0.20718861	0.93751442	0.76038873	0.12470287
0.90471780	0.30598104	0.69342530	0.40672266	0.19950771
0.53654194	0.42368221	0.71321595	0.46615565	0.37798977
0.07253742	0.03331675	0.54706323	0.98252904	0.97160447
0.98686540	0.17675245	0.17872536	0.48158014	0.28095257
0.35149419	0.58039200	0.31890452	0.68989861	0.26925039
0.40641463	0.01523419	0.43367243	0.46492743	0.88651228
0.13472629	0.82974660	0.76594281	0.12793708	0.87413740
0.09339005	0.69310355	0.31811082	0.67073286	0.16139901
0.93179834	0.13819921	0.44300961	0.41426468	0.49850106
0.26262355	0.08923191	0.17177904	0.22758687	0.81950998
0.86877787	0.83707714	0.20346177	0.68707621	0.29130077
0.56411850	0.76300371	0.50095570	0.13870096	0.32360411
0.69331610	0.24745941	0.24491119	0.24233270	0.24979496
0.31777501	0.65849471	0.09099340	0.61950743	0.89810359
0.81305468	0.79539585	0.45488286	0.57073390	0.33045733
0.84613895	0.10271746	0.00105406	0.08186698	0.48171544
0.15349007	0.58550072	0.13159370	0.52005577	0.93599129
0.93544579	0.18875337	0.71350765	0.58226502	0.07202107
0.19174075	0.50225484	0.28786218	0.20687902	0.65051401

FIGURE 8.14. Pseudo-Random Numbers Generated by a Computer Library Program

In this book, we are assuming that a simulation user has available a satisfactory uniform random number list or generator. Figure 8.14 shows uniform pseudo-random numbers obtained from a computer center generator using parameters supplied by the author. Sometimes, computing centers supply random numbers from distributions other than the uniform distribution. For example, a user may have the option of drawing random numbers from, say, the distributions shown in (h), (i), and (j) in Figure 8.13. To distinguish these from uniform random numbers, they are sometimes called *random variates.* Usually, the user can supply parameters that govern the variation by telling the generator the general magnitude of the numbers and the shape of the distribution.

Next Event Simulation

So far in our models, time has been represented by cycles through the model. Each cycle represents either an entire period of time or its ending point. This is *fixed time increment simulation.* As the model operates, it may be that nothing occurs during a time period. In this case, the model cycles uselessly. To avoid this problem, a technique known as *next event simulation* has been developed.

In next event simulation, increments in simulated time can vary in a way that permits the model to move immediately to the next point in time when something will occur. It does this without repeatedly cycling through the model until something happens. For next event simulation, a representation of the continuum of simulated time is required. The point representing simulated present time is a variable usually called "clock." At the beginning of a simulation, the clock is set to zero. The simulation run terminates when the clock reaches a preset ending value of simulated time, when an event counter reaches a set maximum, or when nothing further is scheduled to happen.

Random variates drawn in timing segments of next event simulation represent intervals of time between events. Moments when events occur can be thought of as points on a straight line representing the time continuum. After points when events occur have been generated and placed on the time line, the clock may be advanced in time directly from event time to event time without having to cycle through fixed intervals on the time line until the next event occurs. Thus, cycles through the model need to be executed only when some significant change in the state variables will result. If the event times on the time continuum represent different kinds of events, the model need be cycled through only the part relevant to the particular event. These two features of next event simulation—moving directly to the next event and executing only the relevant part of the model—can result in important reductions in real run time compared to fixed time increment simulation.

The classic illustration of next event simulation is a model of a waiting line object system. Waiting line simulation models generate random intervals between arrivals and random intervals of time required to service each "customer" after arrival. The points in time when arrivals occur are placed on a time continuum. Service ending times are added later depending on when service starts. In execution, the model moves to the next event to occur. If it is an arrival, it executes the arrival segment of the model; if it is the end of a service time, the model moves to a service segment. Next event simulation of a waiting line problem is illustrated by flow diagram in Figure 8.15 for an open-shop man-computer version of the bidding game of Exercise 6.8.

In fixed time increment simulation, cycles represent points in time that are separated by equal intervals. A state history may be recorded at each of these points. In next event simulation, if a state history is recorded only whenever an event occurs, the time intervals between recorded states will be unequal. If equal intervals of simulated time are desired between recorded states in next event simulation, equal interval points on the time continuum must be established for which state variables are to be recorded. With this device, recording a state history is merely execution of one of a special set of events inserted at equal time intervals into the stream of events.

Complex Models

We have already learned the moral: never simulate when analysis will serve. Simulation has become popular and worth the effort because most object systems of practical interest are so complex that models of them cannot be solved directly by analysis.

However, there is another reason for the popularity of simulation. This reason occurs in situations in which analysis would serve, but simulation is easier. The models in these situations contain a large number of independent discrete probability distributions. For models of this type an overall total combined probability distribution (called a joint probability distribution) could be obtained by straightforward but overwhelming computation using the addition and multiplication rules of probability theory. Analytical formulas have been developed for some complex joint distributions, and for some of these the distributions have been tabulated so that the work of computation once performed need not be repeated.

The binomial distribution found in most probability and statistics books is one of these tabulated distributions. Graphic examples are shown in Figures 8.8, 8.9, and 8.13. We can use tables of the binomial distribution to find the number of heads to be expected when tossing a given number of coins.

The binomial distribution can be applied to both fair coins and to bent coins, but not to combinations of fair and bent coins. East separate coin tossed is a separate chance process which is represented in a model by an independent stochastic process. One compound event from the joint distribution of all the independent stochastic processes is a list of the coins on which heads appeared. This list could be obtained by a separate Monte Carlo simulation of each coin. The compound event, which would be the number of heads that appeared, would represent one trial in an experiment. Repetition of this Monte Carlo experiment would generate a distribution of compound events that could be taken to approximate the joint distribution to be obtained by analysis. When analytical formulas and tabulated distributions are not available, it is frequently easier to approximate the analytical result by Monte Carlo simulation than to find the exact answer by "brute force" computation.

In situations where analysis may be available and, if so, would be easier than simulation, a Monte Carlo simulation may still be used to check the overall behavior and conclusions of the analytical model. This is done by Monte Carlo simulations of the elementary stochastic processes to generate a distribution of compound or ending events. If the simulations do not approximate the analytical conclusions, the method of analysis may be incorrect. Once Monte Carlo confirmation of analysis is obtained, the simulation need not be repeated.

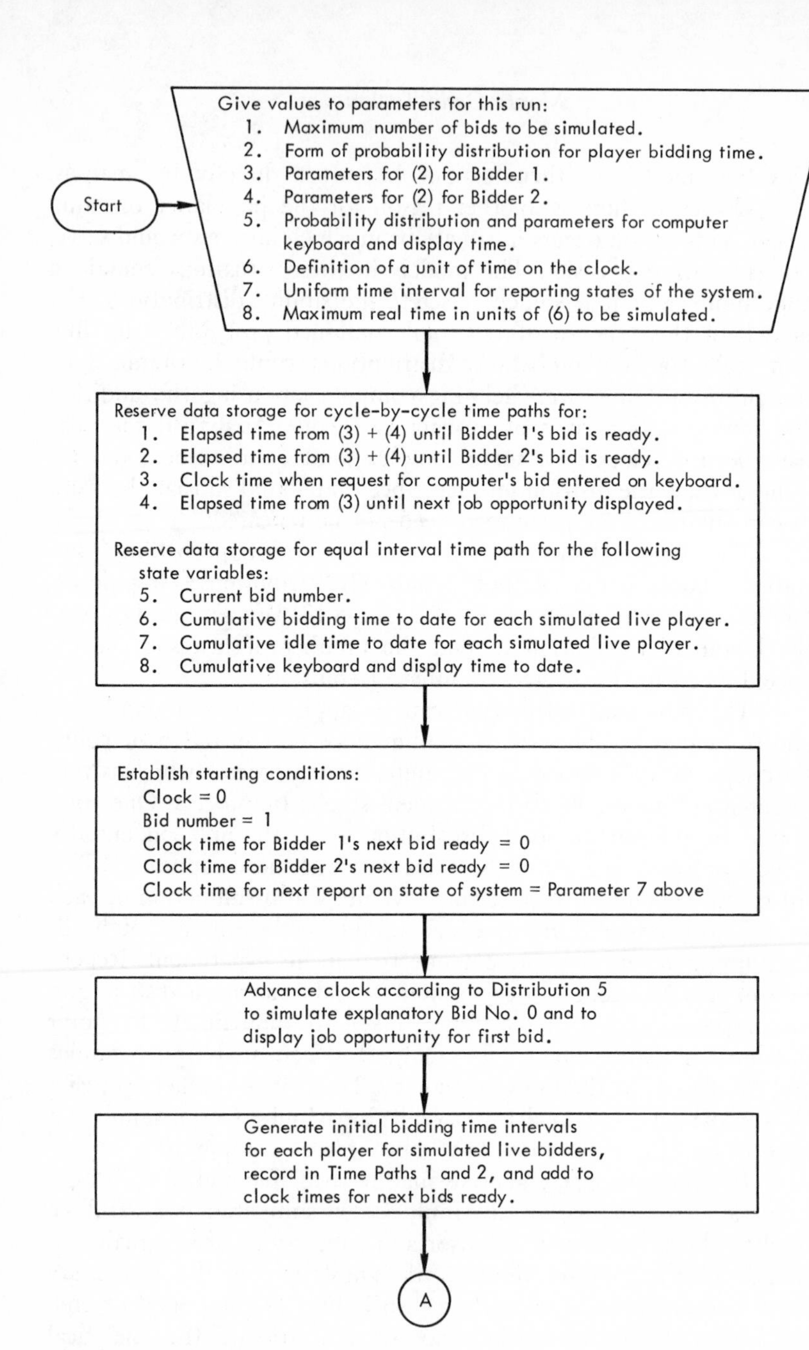

FIGURE 8.15. Next Event All-Computer Simulation of an Open-Shop Man-Computer Bidding Game (One Industry)

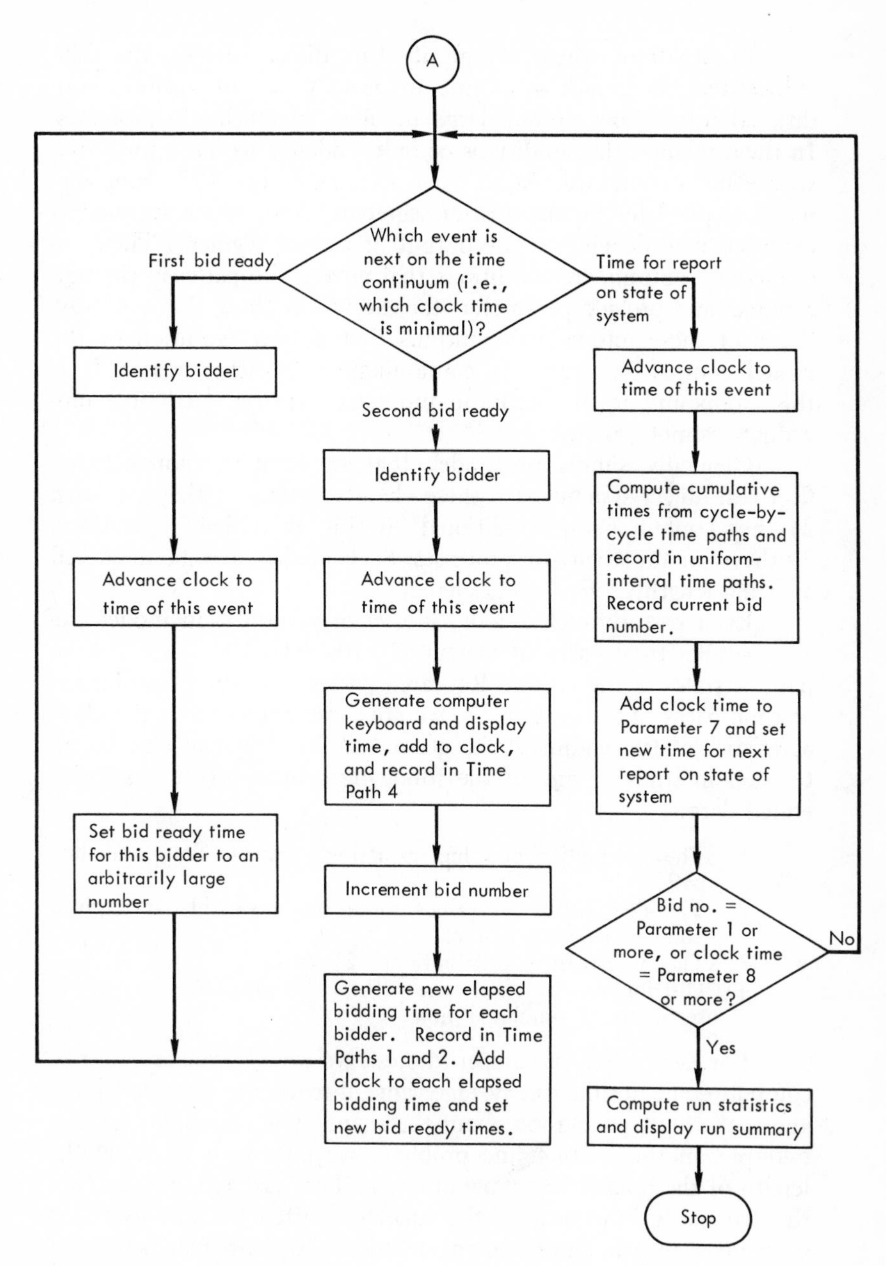

FIGURE 8.15. (*Continued*)

In situations where complexity bars direct analysis, the difficulties analysis cannot surmount are usually due to intricate conditional interactions among large numbers of stochastic processes. In these models, the conditions of independence required for analysis are not maintained. As we have seen in Figure 8.15, some segments depend for input on prior segments. Also, which segment is executed may depend on the output of a prior segment. These intermediate outputs are conditions that govern the pathway through a model and provide parameters for executions along that pathway. Some of these intermediate outputs that govern execution of the model come from segments containing stochastic processes. It is this compounding of stochastic processes that the analytical procedures cannot handle.

Generally, simulation models that are complex enough to reflect real object systems have these characteristics: 1) they are long; 2) they contain many conditional "if this, then that" operations; 3) they contain stochastic processes. Such models may be described as being lengthy, "iffy," and "chancy."

Each pass through a long, iffy, chancy simulation model creates parallel time paths of generated variables which form a time path of states of the system. Repeated passes generate a distribution of time paths of states. Usually, we are most interested in the characteristics of the ending states generated, but we may also be interested in studying any of the following subsets of the total generated data:

1. The distribution of values of a single variable along one time path.
2. The distribution of values for a single variable at a given point in simulated time.
3. The time path of distributions (2) above.
4. Distributions (1) and time paths (3) above, but for separate segments of simulated time.

For each subset of data of interest, we would probably want to compute some meaningful average value, discover the extreme values, and make some conclusion about the shape of the distributions. For example, in the waiting line problem, we may wish to know the length of the longest line; how often the line was zero, one, or two; the average waiting time in the line; how often the line exceeded some given length; and these characteristics for particular periods of simulated time.

A Monte Carlo simulation is really one giant compound stochastic process. As such it generates huge amounts of data that usually must be studied by data reduction techniques. If in the course of studying initial runs, it is discovered that some distributions take on the characteristics of well formulated analytical models, analysis may be substituted for simulation for that portion of the model generating these characteristics. Usually, such substitutions save computer time. However, the substitution of analysis for simulation always loses representation of the dynamics of the object system.

Reliability, Real Run Time, and Cost

If a Monte Carlo simulation model is executed only once, only one value for each variable at each state is generated. For any one state, there is no distribution of values that can be inspected. Over simulated time, the single execution represents only one of the many possible pathways through the model. Thus, for a single execution, we have no way to tell whether the result is an extreme case or about average. Since we are simulating because the model is too complex for analysis, we also have no basis for relying on the single result as a typical output of the model.

Repeated executions of the simulation provide distributions of values at each state and a distribution of pathways. From these distributions we can infer what the averages, the extremes, and the shapes of the distributions would be if we had executed the model infinitely many times, which we cannot do because we stop sometime in order to use our results.

How reliable are the estimates we make from a limited number of executions? Two executions may give us a great deal more information than a single execution. One hundred executions may give us hints of the extremes and of the shapes of the ultimate distributions. A thousand executions may perhaps give us results that approximate smooth curves. But a second thousand executions may suggest smooth curves of different locations and shapes. A third thousand executions may generate results that look even different. On the other hand, the separate sets of one thousand executions may look so similar we would feel no need for further executions.

These possibilities raise the very practical question: How long to run a simulation model? No satisfactory answer has been agreed upon in the literature of Monte Carlo simulation. This question is similar to the one of how large a sample to take. Despite formal theory, a great deal is left to the judgment of the user. Monte Carlo simulation is essentially sampling from known distributions, but complicated by the fact that the distributions from which samples are being drawn are not independent. Periodicities are usually evident in simulation generated data. A statistical technique called spectral analysis is appropriate for dependent data and has been applied to simulation results, yet uniform practical rules for stopping complex simulation models have not developed.

Another important question is: When does "simulation" start? Because of sequential dependence, a "warm up" period is usually recommended. Thus we have the double unresolved questions: How long to warm up a simulation model? How long to run it after it is warmed up?

One answer to these questions is to run it longer than it would ever be thought necessary, then study the results in segments of simulated time. The warm up period would then be determined after-the-fact. Reliability would be judged from comparability of the segments. This suggestion leads to the further question: How long should a segment be? The situation is further complicated by the fact that the warm up for each succeeding segment is the prior segment.

Another answer to these perplexing questions is to stop simulating when the allocated computer time runs out. In this situation, the question of reliability is resolved by obtaining as much reliability (in terms of numbers of executions of the model) as resources will permit. But this answer is not helpful when resources are scarce and more than one simulation project is underway. To know how to allocate resources among projects, one must know the additional reliability gained for the additional resources expended. When the stochastic processes of the model are entirely independent, reliability increases less than proportionately to the number of executions. Sequential dependence increases the gain in reliability received for a given increment in executions, but the gain is still not proportionate. The general rule is decreasing returns to reliability for successive equal inputs of executions.

Scarce resources have led to economizing schemes for Monte Carlo simulation. Most of these are called *variance reduction techniques*.

One variance reduction strategy is to run new versions of the model (with changed operations or changed parameters) on identical lists of random numbers. Thus, the runs become directly comparable and variation between runs due to random numbers is absent. This technique is analogous to historical simulations on identical historical inputs.

Another variance reduction scheme is known as the method of antithetical variates. This strategy attempts to balance the random draws along the unit length against each other in pairs. Every second random number is taken to be the difference between the prior number and the number one. Thus the random number .25 would be followed by .75; .5 by .5; .1 by .9; .6 by .4; and so on. This method deliberately introduces a dependence into the sequence of uniformly distributed random numbers. Its purpose is to reduce the spread of the distributions generated by the model. For given reliability, reduced spread means real run time can be reduced. Before using antithetical random variates, a user would want to satisfy himself that the implied "alternating" behavior of his stochastic processes does not invalidate his model.

Simulation prospers despite these unresolved issues. If the operations of a model and its basic stochastic processes are believed to be valid, then any particular result generated by a single pass through the model is a time path that might occur in the object system. The practical question for the user is: How many of these individual possible results should he look at for the purposes at hand. Maybe only a few executions are required. If the first results appear disastrous, the user may decide to avoid completely the object system represented. Problems of trade-offs of resources for reliability are appropriate only when alternative object systems compete very closely through their simulation models for adoption. In these cases, the final decisions may depend solely on the reliability obtained from each model, and additional real run time and other costs may be justified. Sometimes, when object systems compete closely, decision makers become indifferent to the simulation results and choose a final alternative on issues that were not built into any of the models. This practice is discouraging to model builders, but should also indicate to them that their models seldom contain all the important factors.

EXERCISES

8.1. Create some physical device to simulate uniformly distributed random digits 0, 1, 2, ..., 9. You may wish to do this by numbering small tokens such as plastic balls, poker chips, or slips of paper that can be drawn from a container. Before each draw, replace the previously drawn token and mix all tokens thoroughly. How many tokens should you use? Another device would be to toss four pennies to simulate four binary digits, then translate the heads appearing into decimal digits. There are many possibilities. Once your physical device is completed and working, draw one hundred random digits. Keep a tally of how many of each value are drawn. Study the final distribution and make a judgment about the uniformity of the output of your physical generator. Use your one hundred random digits to plot fifty points in the unit square. What does this tell you?

8.2. Assume you have a uniform random generator of decimal digits. Describe as many schemes as you can think of to draw random decimal digits to simulate (a) the tossing of two fair coins, and (b) the tossing of a fair coin and a bent coin. For each scheme, state how you would make some judgment about the fairness of your fair coins.

8.3. Use your physical random decimal digit generator of Exercise 8.1 to simulate the chance processes of Exercise 8.2 for a "reasonable" number of trials. Try to use all the different schemes you thought of for Exercise 8.2. Judge the fairness of your simulated fair coins.

8.4. Describe as many different schemes as you can think of to use random decimal digits to simulate (a) the tossing of a single die and (b) the tossing of a pair of dice.

8.5. Use your physical random digit generator of Exercise 8.1 to execute the various schemes you thought of for Exercise 8.4. Simulate at least sixty tosses under each scheme, keep a record of results, and make judgments about your schemes and about your random digit generator.

8.6. Develop a scheme using random decimal digits to simulate the tossing of two coins, the first of which is a fair coin and the second of which is either a fair coin or a bent coin, depending on the outcome of the first coin. Draw the analytical tree diagram for this object system (see Figure 8.3). Compute by analysis using the multiplication and addition rules of probability theory the probability distribution for the compound events. Use your physical random decimal digit generator of Exercise 8.1 to carry out one hundred executions of this process. Compare your results with the results expected by analysis. Make some judgment either about the adequacy of analysis or of the adequacy of simulation.

8.7. Write a short philosophical essay on the issues involved in making the judgment asked for in Exercise 8.6.

8.8. Create from your imagination an unusual (and hopefully unique) game of chance. Let this game involve tossing coins and dice. The coins may be fair or bent; the dice may be fair or biased. Have your game involve at least six stages. Permit the rules to declare winners at any stage, or to send the player back to start all over again from any stage. At each stage, allow alternative chance processes, but only one is to be executed at a stage. Which chance prcoess is to be executed should depend on inputs or intermediate outputs up to that stage. Draw a flow diagram of the game you have imagined.

8.9. Develop a scheme for most efficient Monte Carlo simulation of your game of the Exercise 8.8.

8.10. Execute your scheme from Exercise 8.9 by using random numbers from Figure 8.14. You may use these random numbers as decimal numbers between zero and one, or as random integers by selecting one or more digits at a time from the digits to the right of the decimal point. Proceed through Figure 8.14 only in a uniform manner determined arbitrarily in advance.

8.11. Design a Monte Carlo simulator of events that are points on a continuum. These may be water levels, speeds, heat levels, times, etc. One possibility is simulating the amount of time a participant in a man-machine simulation requires to make his response on a remote terminal after seeing the stimulus displayed. Draw a curve representing the distribution of probability over event points (the area under this curve should be one). Then draw the cumulative probability curve for this distribution. Use your simulator graphically with one hundred random numbers (generated in any convenient way). Mark each point-event on the horizontal axis and make a visual comparison of the density of points along the line with the height of your curve representing the distribution of probability over even event points. Make some judgment about your Monte Carlo distributions.

8.12. Design a next event simulation of the man-computer simulation of Figure 6.4 to determine how much real computer time would be required. Assume the times required by the computer are fixed so that the only variable in determining the total real time required is the behavior of live participants. Assume that it takes one minute of real computer time to execute the computer simulation program of Figure 6.4, and ten seconds of real computer time to add a response to the data base. Assume also that responses are not obtained in a conversational mode so that the physical act of responding is always the same. Thus, the only time varying phenomenon is the "thinking" time used by the participants. Prepare a flow diagram of your design. You may hypothesize the distribution of times required for live responses or use your point-event simulator from Exercise 8.11.

8.13. Consider as an object system a man-computer simulation of the example game on an open-shop dedicated computer for the case where one live participant makes a conscious deliberate decision and the other player is simulated in the computer. Assume the dedicated computer is "on the air" for this purpose only between 2:00 p.m. and 4:00 p.m. on Thursday, and that an instructor will tell his class of forty students to individually drop by the dedicated computer and play one 101-trial run of the example game. Hypothesize a discrete distribution of times between arrivals during the two-hour period. Hypotheize a discrete distribution of times required to service each student after his arrival. Draw a flow diagram of an all-computer simulation model to determine whether the computer two-hour period should be lengthened or shortened. Provide for reports from your simulation of the length of the line of waiting students, the average waiting time in the line, and other interesting statistics.

8.14. Using any convenient source of random numbers, simulate the 2:00 p.m. to 4:00 p.m. Thursday behavior of students at the dedicated computer in Exercise 8.13. Should the period be shortened or lengthened?

8.15. Exercises 8.12 and 8.13 have asked you to design simulations of simulations. You could, of course, continue this regression through many simulations of simulations of simulations. Write a short essay on the philosophical issues these situations create.

9

SIMULATION LANGUAGES

The advent of electronic digital computers has brought with it hundreds of new languages. Some of these languages refer to the actual operations computers perform and thus provide part of users' conceptions of what computers are and of what they can do. Other languages refer to general areas of computer applications. When used to instruct computers how to solve problems, these languages force users to see some aspects of their problems in certain ways. Such influences on users' conceptions of computers and on their views of problems appear in varying degrees in all computer languages.

Usually, by the word *language* we mean an enduring system of communication that evolves within a community of persons. New knowledge and new activities do not induce new languages, but are usually articulated in an existing language, such as English or mathematics. Although new words are invented and new meanings evolve within the community using a given language, seldom is a new grammar created or is syntax altered. Thus man-model and man-machine simulation (where the machine is not an electronic digital computer) have not required new or special languages. Also, we have not needed new languages to operate many kinds of machines. For example, "plus" and "minus" on desk calculators are symbols in the existing mathematical language. To drive an automobile, we may learn some new terms and meanings, but we do not need a new language.

Then why should electronic digital computers cause the birth of hundreds of new languages, each complete with an individual syntax of its own?

The obvious answer is that computers are not human; they cannot learn or adapt to the one language of the whole community of computers, if there were but one computer language. The more technical answer is that electronic computers "talk" to themselves, but each in a unique way. To carry on this internal "conversation," digital computers store programs and data within their memories and then proceed to "tell" themselves what to do next. It is as if they stored sentences, paragraphs, chapters—even complete books—in their "minds," and then read through them word by word. To store sentences, paragraphs, chapters, and books somewhere, one needs a language. As computer technology developed, neither English nor mathematics was a sufficient language in which to write the "book" of programs and data stored internally by computers. As a result, man now designs rather than evolves new languages, at least where computers are concerned. And he has designed a great many.

In this chapter, we shall discuss languages that have been specially developed for writing computer simulation programs. First, let us define five levels of computer languages:

Level 1. Internal machine language.
Level 2. External machine language.
Level 3. Machine-oriented symbolic language.
Level 4. User- or problem-oriented language.
Level 5. User-written language.

These levels may be interrelated in a single application. For example, a user may write his own special language (Level 5) in a user-oriented language (Level 4) that is translated into a machine-oriented symbolic language (Level 3), which in turn is translated into external machine instructions (Level 2) that when read into the computer become magnetically stored in the computer's internal language (Level 1). The amazing thing about internally programmed computers is that they perform these translations from one language to another themselves.

The Hierarchy of Computer Languages

Languages for controlling digital computer operations are appearing with bewildering frequency. They are identified by acronyms such as MAP, MAD, SPS, ALGOL, FORTRAN, COBOL, PL/I, SLIP, LISP, QUICKTRAN, JOVIAL, BASIC, and SNOBOL. Some of these languages are in their third or fourth dialects. Later in this chapter, we shall illustrate GPSS, SIMSCRIPT, and DYNAMO.

First, we shall describe the language levels listed above. To illustrate these levels, we shall use an example set of computer instructions. These instructions are part of a program segment to simulate coin flipping. Given a decimal random number between zero and one, they test whether it equals or exceeds one-half. If so, the result is declared to be a head, which is recorded by storing a "one" in the storage location for the variable named RESULT. If not, the "zero" recorded in RESULT before the test remains and represents a tail. This set of instructions, written in FORTRAN IV, is as follows:

```
TAILS  = 0.
HEADS  = 1.
HALF   = .5
RESULT = 0.
IF (RANDOM . GE . HALF) RESULT = HEADS
```

LEVEL 1. INTERNAL MACHINE LANGUAGE. Ultimately, all computer languages are translated into a string of binary digits called "bits." Bits are physical representations that can take on two possible states. Examples are: areas on a card or on paper tape that may be punched or not with a rectangular or round hole; lights that may be switched on or off; blank areas on paper that may be marked or not with a pencil; spots on tapes, drums, or discs that may be magnetically coded in one of two ways; tiny interconnected doughnut-shaped ferrite cores, or tiny rods, that may be magnetized in one direction or in the opposite direction; or simply a string of zeros and ones written on a piece of paper.

Digital computers read bit strings either as instructions or as data. Individual instructions are represented by distinctive strings of bit values of a given length. The instructions themselves are carried out by the circuitry of the machine. After an instruction is executed, the computer automatically proceeds to the next instruction without human operator intervention. So to speak, the computer is reading a procedure manual which is written in a language that has an alphabet containing only two characters. The lights seen flashing on and off on computer consoles are "speaking" the two-character internal machine language. The total number of words in this language is established by the computer manufacturer, and is called the *instruction set.* A program in machine executable form is called an *object program.* An internal machine language representation of part of the example instructions to create the result heads or tails from a random number is as follows:

```
0100 1110 0000 0000 1101 0000 0101 1100
0100 0110 0000 0000 1101 0000 0100 0000
0100 1110 0000 0000 1101 0000 0011 0010
0100 0110 0000 0000 1101 0000 0100 0110
0100 1110 0000 0000 1101 0000 0011 1010
0100 1111 0000 0000 1101 0000 0011 0110
0010 1111 0010 1000 1101 0000 0100 0000
0100 0110 0000 0000 1101 0000 0100 0110 [1]
```

LEVEL 2. EXTERNAL MACHINE LANGUAGE. Computers can read punched cards and paper tape, and magnetized surfaces such as those on tapes, discs, and drums. Sometimes what is on these external storage media are duplications of internal machine language. Reading this type of input is very efficient for the machine because no translation is required before recording the input in main storage. However, from the point of view of the user, internal machine language is inefficient because to write a pro-

[1] Eight-bit code translation of IBM/360 hexidecimal listing of selected instructions.

gram he must translate his model into a long string of bits using a lexicon made up of only two characters. To assist machine-language programmers, computer manufacturers describe machine instructions in ways that aid human memory. They do this by representing machine instructions by numbers or alphabetical letters. These representations form a language that is simply the instruction set for the machine stated in numerical or mnemonic symbols rather than in the bit strings of internal machine language. Thus, external machine language corresponds instruction for instruction with internal machine language. The syntax for both internal and external machine language is provided by the design of the machine itself.

Alphabetical characters, decimal digits, and other special characters are usually represented on external media such as punch cards in codes different from internal machine language. Some early computers recorded characters internally as they were coded externally. This is now rare, and most present computers contain circuitry that translates characters from one code on external media into correct strings of bit values for internal storage. For example, your name recorded on a familiar punched card would show two punches in each column. Computer hardware on reading this card would translate each column into a string of bits of fixed length in computer storage to represent each different alphabetical letter in your name. These would then be stored as data in computer memory. Similarly, external machine language (usually in its numerical rather than mnemonic representation) is translated into internal bit strings that are computer instructions. External machine language for the previous part of the example program segment to test a random number to determine heads or tails is shown below:

```
78  00  D  05C
70  00  D  064
78  00  D  050
70  00  D  070
78  00  D  058
79  00  D  054
47  00  D  0A4
78  00  D  064
70  00  D  070
```
[2]

[2] IBM/360 hexadecimal listing of selected instructions following FORTRAN IV H compilation.

LEVEL 3. MACHINE-ORIENTED SYMBOLIC LANGUAGE. Internal computer instructions that operate with internally stored data or that cause input-output devices to perform specified functions must include the location (called the *address*) of the internally stored data required or the identity of the device to be actuated. Thus, an internal instruction contains an operation code plus addresses and/or device identifiers.

Historically, one of the first breakthroughs in making computers easier for human programmers to use was the development of symbolic languages. Not only were operation codes represented by external alphabetical mnemonic symbols, but the addresses and device identifiers were also symbolically represented. A computer program, nowadays called an *assembler,* translates programs written in symbolic languages by assigning internal machine addresses or external device identifiers to the external symbolic names used by the programmer. These are then associated with the proper internal operation code, which the assembler also translates from a mnemonic representation.

The next step in language development was grouping several machine instructions together and representing them symbolically by a single master instruction. The reason for doing this is that individual instructions may be written hundreds of times in a single program. Once computer tasks that require several machine instructions are defined, they may be represented by symbols or words called *macros.* A *macro* is a labeled set of machine instructions to perform a specific task. Macros are written either in external machine language or in the symbolic language of which they later become a part. Each macro is given a name and a machine-readable label. Translators interpret each macro label as it is read and create from it the specific set of machine instructions it represents. Macro translators are usually incorporated in assemblers. Symbolic machine and macro languages at this level are frequently called *assembly languages.* The translator or assembler, which is itself a computer program, may even have been written in part in the macro language it translates.

Because translation of symbolic or macro language statements is the execution of a computer program, the new program that the translation process creates must be stored elsewhere until the translator program releases control of the computer. Subsequently, the new program can be read into the computer's main storage and executed. We have now reached a level in the hierarchy of languages where we see computers writing programs for themselves. This self-programming feature provides a dramatic saving in programming time for human programmers.

Because assembly languages are related to machine languages, they are oriented to particular computers. Thus, each computer has a unique assembly language of its own, just as it has a unique instruction set. If users who write at the symbolic- or assembly-language level change computers, it may be necessary for them to learn a whole new language. Part of the example segment to test a random number to simulate the result heads or tails is shown below in a symbolic language:

```
MVI   99(13),X'02'
MVC   HALF(4),C..03DC
MVI   99(13),X'03'
MVC   RANDOM(4),C..0420
MVI   99(13),X'04'
MVC   HEADS(4),C..0454
MVI   99(13),X'05'
MVC   TAILS(4),C..04A4
MVC   RESULT(4),C..04A4
MVI   99(13),X'06'
LE    O,RANDOM
CE    O,HALF
BC    4,CL.1
MVI   99(13),X'07'
MVC   RESULT(4),HEADS [3]
```

LEVEL 4. USER- OR PROBLEM-ORIENTED LANGUAGES. Machine and symbolic languages are specific to particular computers. Most computer users are more interested in their own problems than in the problems the computer has in providing service. For them, languages oriented to their own backgrounds and own problems are more desirable. Such languages may look

[3] Mnemonic listing of selected instructions following IBM/360 PL/I F compilation.

much like languages users already know. For example, FORTRAN —the acronym comes from FORmula TRANslation—is a language that looks much like algebra.

Also desirable are languages that computers can translate but that are not specific to any particular make or model of computer. A language that is common to many computers saves users the necessity of learning new languages when their computer facilities change. Some users may have access to computers of different kinds, and problems written in a common language can then run on all of them without reprogramming. Where computer programs are written by teams, a language common to team members as well as to the various computers they may use is a necessity. Also, as computer hardware moves through generations of innovation, common languages can remain much the same thereby giving users the benefits of new computer technology without abandoning their learning investments in computer languages. To preserve the commonality of languages over time and for different implementations, official standards have been set or proposed for some computer languages.

If a language is to remain relatively constant as computers change, then a translator specific to each computer must be created so each different computer can understand the common language. Machine-specific translators to do this are called *compilers*, and the process of translation is called *compiling*. Compilers translate programs written in user-oriented languages either directly into internal machine language or into symbolic language instructions, which in turn are translated into internal machine language by an assembler. Compilers and assemblers are themselves only computer programs. In the translation process, programs written in higher level languages called *source programs* are changed into equivalent machine codes called *object programs*. Translation begins in a *source language* and ends in an object or *target language*. Again, we should note the remarkable ability of stored-program computers to program themselves. A recent innovation is the "forgiving" compiler that allows users to make (limited) mistakes.

An ideal language would be common to all computers and users. This ideal has not been achieved. Most source languages are oriented to particular kinds of problems. Later in this chapter, we will illustrate in detail three source languages developed for simulation. Each implementation of a source language on a particular computer usually imposes a few specifics due to the uniqueness of the machine. The computer software industry has now progressed to the point where most computer users need learn only one or two source languages. The heads-tails segment at the beginning of this section was written in FORTRAN IV. This same segment written in PL/I (Programming Language One) is shown below:

```
HALF = .5;
RANDOM = .75;
HEADS = 1.;
TAILS, RESULT = 0.;
IF RANDOM > = HALF THEN RESULT = HEADS;
```

LEVEL 5. USER-WRITTEN LANGUAGES. Users who know only user-oriented common languages may themselves wish to develop new application languages. This may be particularly true for man-computer simulation designers and for writers of computer-assisted instruction programs. To these users, the task of learning one or more machine specific languages may be so formidable as to alter or stop their projects. Fortunately, it is possible to write a new vocabulary and a new syntax using as a basis one of the common user-oriented languages. Thus is developed a kind of "super" level of languages that are originally written in user-oriented languages.

Programs written in user-written languages may become executable object programs in two ways, depending on the implementation. First, the capabilities of a Level 4 language may be programmed directly to process new words and new syntax pertinent to a particular application. An example is a language for conversation between computer and subject which is written and executed in a Level 4 computer language. Second, programs in user-written language may be translated by a special computer program into a Level 4 language, which is then compiled as usual. This second way requires an additional machine translation step. Also, the language writer must either write his translator program or spend resources to have a translator program written for him. Original SIMSCRIPT is a Level 5 language that is in turn translated into FORTRAN (a Level 4 language) that is in turn compiled into machine executable instructions (Level 1). The coin-tossing example written in original SIMSCRIPT is:

```
LET HALF = .5
LET HEADS = 1.
LET TAILS = 0.
STORE TAILS IN RESULT
IF (RANDOM) GE (HALF) STORE HEADS IN RESULT
```

One way to write a new language directly in a Level 4 language without an intervening additional translation is to write a set of separate computer routines. Each of these routines would have a name and would perform a specific task. Given this set of names as a lexicon, and the interrelationships of these tasks as a syntax, the user can write master or control programs called *main programs* for different applications simply by writing the names of the several routines in the order he wishes and with the inputs he desires. To illustrate, imagine the FORTRAN IV heads-tails example above is a routine named FLIP. The calling statement may appear as follows:

```
IF (EVENT . EQ . TOSS) CALL FLIP (RANDOM, RESULT)
```

The Level 5 language in this statement is EVENT, TOSS, FLIP, and RESULT. If the next event is to be a toss, the Level 4 statement above would send to the routine a random number and return to the main program a zero or one in the internal storage location called RESULT.

The Variety of Simulation Languages

In the computer milieu, the number of simulation languages is growing rapidly. Some simulation languages are written for particular applications such as man-machine simulation of complex organizations. Some may also be used for nonsimulation purposes. Of course, simulations may be programmed in any computer language. Special languages designed for purposes other than simulation, particularly those designed for computer assisted instruction, may prove to be convenient for simulation programming.

Most of these special languages are Level 4 and Level 5 languages. Later in the chapter, we shall review three of these languages and investigate the ways of viewing problems they impose upon their users. Before doing this, let us scan an annotated list of other simulation languages.

1. *FLOW DIAGRAMS*

In this book, we have used flow diagrams as a language. The shapes of the figures represent new words and the rules for following solid and dashed lines represent a new syntax. These features are combined with mathematics and English to form the complete flow diagram language. A flow diagram is a practical means for writing the logic of a simulation whether the simulation be man-model, man-computer, or all-computer. What is shown in flow diagrams can also be shown in ordinary prose, but without the convenience and visual appeal of flow diagrams. Except in special applications, such as GPSS discussed later, flow diagrams are not a sufficient language for communicating instructions to digital computers.

2. *DECISION TABLES*

In flow diagrams, the diamond symbol represents testing of conditions. Depending on a test result, one of as many as three different pathways may be followed. By combining diamonds, complex conditions and multiple pathways may be stated; however, the resulting flow diagram may become unwieldy. In contrast, decision tables are a compact way of stating both complex conditions and the complete set of operations to be performed for each possible combination of these conditions. In decision tables, both the conditions and the operations may be stated in English. These are listed in a tabular form that makes it easy to read what operations are to be performed under any combination of the presence of, absence of, or indifference to the conditions. Everything stated in flow diagrams can be stated in decision tables and *vice versa.* The advantage of decision tables is conciseness. Like flow diagrams, decision tables are not a sufficient language for writing computer programs. Both must be translated to a computer-understandable language by a human programmer. Attempts have been made to write computer programs that will translate decision tables into object programs, but the results of these efforts are not yet generally evaluated.

3. *QUESTIONNAIRES*

Attempts have been made to reduce computer programming to the task of filling in a questionnaire. For example, the RAND Corporation developed a job shop simulation program generator that runs on the answers supplied by users on a multiple-choice English language questionnaire.[4] The questionnaires are oriented for persons who know nothing about computers. It has been suggested that the ultimate in programming by questionnaire would occur when users would sit at remote terminals merely replying *yes* or *no* to questions asked by the computer.

4. *ONLINE COMPUTATION AND SIMULATION (OPS-3)*[5]

This language is really an entire system for online computation and simulation. It developed out of an experiment in a graduate seminar on advanced computer systems at Massachusetts Institute of Technology. OPS-3 is a hybrid Level 5 language embedded in the M.I.T. time sharing supervisor language. OPS-3 provides man-machine interaction for the simulation designer. This is what we have called in this book online real-time all-computer simulation.

5. *ENVIRONMENTAL CONTROL LANGUAGE-1 (ECL-1)*[6]

This language is specific to a particular computer in a particular place: the computerized laboratory for management science of the Center for Research in Management Science at the University of California, Berkeley. It is specifically designed for man-computer simulation with multiple subjects using a dedicated computer. Again, this language is an entire man-computer system enabling

[4] Allen S. Ginsberg, Harry M. Markowitz, and Paula M. Oldfather, "Programming by Questionnaire," in *Digital Simulation in Operational Research*, ed. S. H. Hollingdale (New York: American Elsevier Publishing Co., 1967), pp. 131–140.

[5] Martin Greenberger et al., *Online Computation and Simulation: The OPS–3 System* (Cambridge: M.I.T. Press, 1965).

[6] Austin Curwood Hoggatt, "Environmental Control Language–1," University of California Center for Research in Management, Working Paper No. 225, mimeographed (Berkeley, 1967).

subjects to interact with each other and with the computer via teletypewriters. It is essentially a Level 3 language very specific to the machine hardware and the laboratory site for which it is written.

6. *TSS AND TINT*[7]

System Development Corporation in Santa Monica, California, is one of the pioneers in time sharing. Its time sharing system and language is called TSS. In this system, users communicate with a central computer from teletypewriters in an interactive online mode. TSS has served users all over the world. TSS is a Level 3 language specific to a particular computing and time-sharing implementation. Embedded within TSS is TINT, an on-line computer interpreted version of JOVIAL, a Level 4 algebraic-like language. (Other languages are also available to the TSS user.) Man-machine simulation users may program in JOVIAL new languages specific to particular man-computer interactive experiments. Computer assisted instructions have also been programmed on this system. These are Level 5 applications of the Level 4 JOVIAL language, which is in turn controlled by the Level 3 TSS system and language. In this computer system, JOVIAL is interpreted statement by statement into symbolic language rather than compiled all at once into executable code.

7. *EXPERIMENTER ORIENTED LANGUAGE (EOL)*[8]

The U.S. Army's Behavioral Science Research Laboratory in Washington, D.C., needed a language in which behavioral experimenters could deal with their subjects without turning themselves into computer programmers. Also, because of the variety of equipment with which subjects can interact in this laboratory, it was desirable to eliminate for experimenters the task of becoming equipment engineers. EOL was written to fill these needs, but it is specific to a particular computer and laboratory. This language "speaks" to experimenters in their own stimulus-response research framework. It provides the means by which the experimenter can

[7] *Command Research Laboratory User's Guide,* System Development Corporation, Document No. TM–1354/002/03 (Santa Monica, Calif., 1965).

[8] "Experimenter Oriented Language," Behavioral Science Research Laboratory, Army Research Office, Washington, D.C., Technical Report No. TR–67–601–02 (Bethesda, Md.: Informatics, 1967).

design the expected responses from subjects, relate these responses to custom programs for his particular experiment, and design the feedback (stimuli) displayed to subjects. The EOL language and system is controlled by a multiprogramming system that allows the computer to perform high speed "background" computing while waiting for human subjects to respond at their relatively slow speed.

8. *SIMULA*[9]

This language is an extension of a Level 4 language, ALGOL 60, and it contains ALGOL 60. Thus SIMULA is a new Level 4 language created from a prior Level 4 language by extension. SIMULA programs look very much like ALGOL 60 programs, but with added words pertinent to waiting-line problems such as "when," "then activate new," and "activity."

9. *INFORMATION PROCESSING LANGUAGE-V (IPL-V)*[10]

This is a Level 4 language developed out of attempts to simulate human problem solving of complex ill-structured problems. It is one of the early languages specifically written for simulation purposes. It is also one of the early languages designed to create, process, and dynamically structure lists. In IPL-V, lists are not stored sequentially within the computer memory. Each item in a list (wherever it is stored) also contains an address telling where to go in computer memory for the next item in the list. This language views human information processing as list-searching activity.

10. *PROGRAMMING LANGUAGE FOR INTERACTIVE TEACHING (PLANIT)*[11]

This language is specifically designed to handle computer-human interaction problems using time sharing systems. Although

[9] Ole-Johan Dahl and Kristen Nygaard, "SIMULA—an ALGOL-Based Simulation Language," *Communications of the ACM*, Vol. 9, No. 9 (September, 1966), pp. 671–678.

[10] Allen Newell et al., *Information Processing Language-V Manual*, 2nd ed. (Englewood Cliffs, N.J.; Prentice-Hall, 1964).

[11] Samuel L. Feingold, "PLANIT, a Flexible Language Designed for Computer-Human Interaction" (1967 Fall Joint Computer Conference), *AFIPS Proceedings*, Vol. 31 (Washington, D.C.: Thompson Book Co., 1967).

the motivation for creating PLANIT was computer-assisted instruction, its features make it applicable to some man-computer simulation projects. Acronyms of other computer assisted instruction languages are PLATO (Program Logic for Automatic Teaching Operation), COURSEWRITER, SOCRATES, MENTOR, and LYRIC.

World Views

How each of us views the world we live in, and find problems in, depends largely on our training and experience. In effect, each of us has a unique pair of glasses (not necessarily rose colored) through which we individually view the world. Our ways of viewing the world act like filters and affect what we see. Sometimes we deliberately choose special kinds of filters for seeing particular problems of the world. For example, we may put on our mathematical glasses when we want to build an analytical model of a problem. In contrast, we may put on human relations glasses (which filter out technical issues) to see clearly the causes of conflicts among people.

Computer languages cause their users to wear very specific glasses through which to see the world. The choice of a computer simulation language is in effect a choice of a world view. A computer language embracing all world views would be so general that any compiler written to implement that language on a computer would be grossly inefficient. This is one of the difficulties with ordinary English as a computer language. Hence, it is for very practical reasons that computer languages, and especially simulation languages, limit themselves to particular kinds of problems and to rather fixed ways of viewing these problems.

In the remainder of this chapter, three simulation languages are illustrated and their world views described. These languages are GPSS III, SIMSCRIPT II, and DYNAMO.

Example Problem

The three languages will be illustrated by writing in each language the same simulation problem. This problem is derived from the all-computer next-event simulation of the open-shop man-computer bidding game shown in Figure 8.15. For our present purposes, we shall modify the problem of Figure 8.15. First, let us simplify this problem by omitting much of the detail. The problem that remains is a service facility (the open-shop computer) that serves customers or users (the two live bidders). The two bidders coupled with the computer form a single bidding industry. Thus, the simplified problem is the computer serving one pair of subjects whenever both subjects are ready with their bids. In this situation, the computer is usually waiting for the live subjects to complete their bids.

Let us now make this simplified problem more complex by adding additional pairs of subjects. Then, let the job opportunities to be bid on be the same for all pairs for identical trial numbers. In effect, each industry (a live pair plus the computer) represents a separate run of the man-computer simulation, but with identical stimuli (the job opportunities). However, the computer can accommodate only one industry at a time. In this situation, pairs of subjects must wait for the computer to become available. They form a waiting line (a queue) of pairs in front of the computer console. Assume the computer keeps track of job opportunities by trial numbers and of industries by identification numbers. Thus, the computer may be addressed at any time for any pair of subjects for any bidding opportunity without regard to the trial number any other pair may be bidding on. This means that some pairs can be fast bidders and finish the specified number of bids early, others can be slower and finish late in the open-shop period, and others may be so slow they do not finish at all. For purposes of our example problem, let us assume the subjects have all arrived on time for pre-session briefing, have been assigned in pairs to numbered industries, and are organized by industry number into an initial queue of pairs ready to approach the computer the moment the open-shop period begins. The job opportunity for the first bid is given during briefing; thereafter, the computer displays the job opportunity for the next trial before the subjects walk away from the console. Let us also assume pairs do not join the queue until their next bids are ready, i.e., subjects do not work on bids while standing in line.

The example problem is now a single service facility serving a finite number of customers who double back once served. The

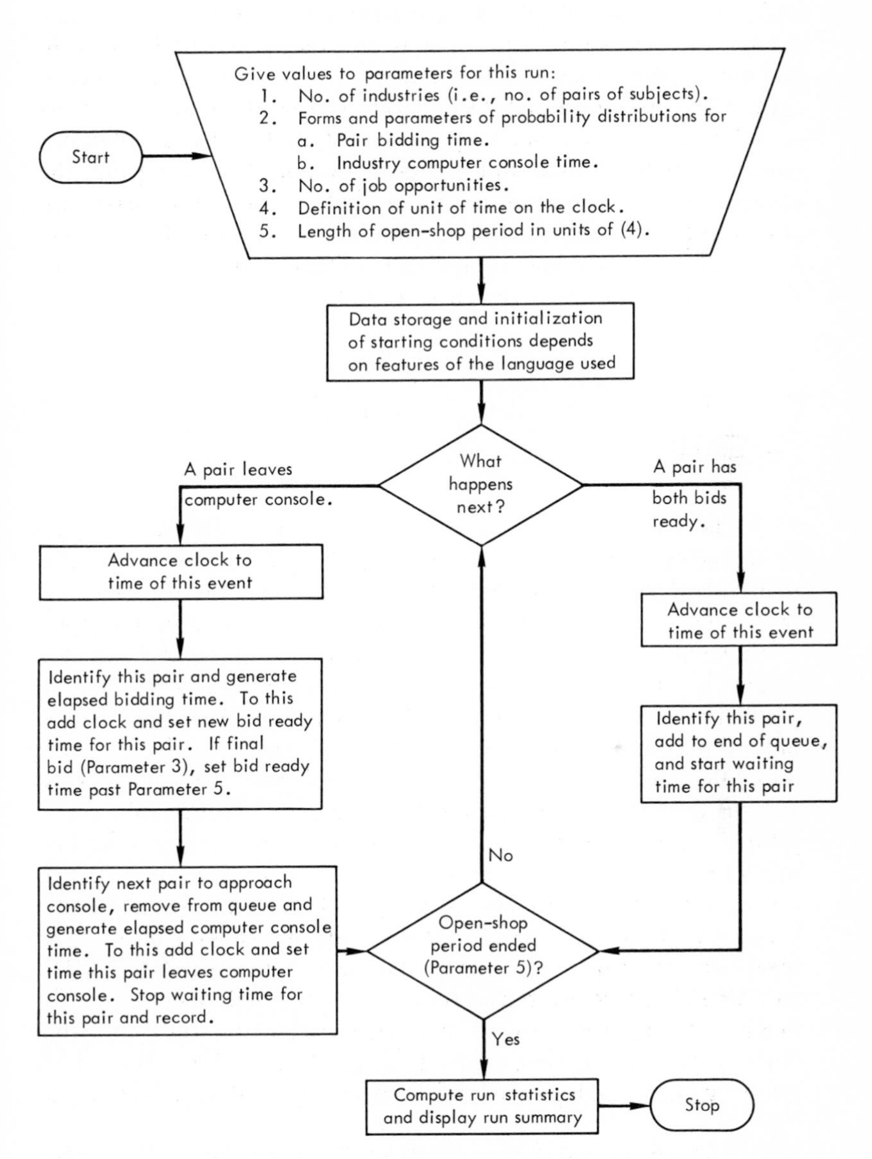

FIGURE 9.1. Next Event All-Computer Simulation of Pairs of Subjects Bidding at the Open-Shop Computer Console

customers are pairs of subjects that require varying amounts of time for bid calculations before rejoining the queue. They also vary in amount of time they use the computer at each trial. Parameters controlling this all-computer simulation will be the number of pairs of subjects, the form and parameters of probability distributions governing bidding time and computer console time for pairs, the number of job opportunities to be bid on, and the maximum amount of simulated real time established by the size of the period for which the open-shop computer is available. A flow diagram of this example problem is shown in Figure 9.1.

General Purpose Simulation System III (GPSS III)[12]

The user who chooses GPSS III as as simulation language simultaneously chooses to view his object system as units of traffic that flow through storages and facilities. The units of traffic are called *transactions*, and they may have a variety of characteristics. Examples of transactions are ships arriving at a port, customers arriving at a theater box office, and pairs of the subjects ready to bid in our example problem. A *facility* is something that is used by transactions, but by only one transaction at a time. Examples of facilities are docks for unloading ships, box offices for the sale of theater tickets, and the open-shop computer programmed to bid against subjects. A *storage* is something that a transaction enters, but unlike a facility, more than one transaction can occupy a storage at a time. When transactions attempt to use facilities that are busy, or to occupy storages that are full, they are delayed and placed in a queue. The measurement of queues is one of the major functions of GPSS III. In GPSS III programs, transactions are created, join queues, wait for facilities or storages, seize them when they are free thus departing from queues, spend time in facilities or storages, leave them thus releasing them, and proceed to the next queue, storage, facility, or

[12] *General Purpose Systems Simulator III, Introduction,* IBM Application Program (White Plains, N.Y., 1965).

leave the system. Sometimes transactions are destroyed within the system and disappear. GPSS III programs do not differ in these general characteristics. They differ only in the pathways that transactions may follow.

To use GPSS III, the user must learn the meaning of such terms as ADVANCE, ASSEMBLE, ASSIGN, CHANGE, DEPART, ENTER, EXECUTE, GATHER, GENERATE, LEAVE, PREEMPT, PRIORITY, QUEUE, RELEASE, RETURN, SEIZE, SPLIT, TRANSFER, and TRACE. There are other terms that deal principally with recording and reporting statistical information about simulations runs. To write a GPSS III program, the user need learn only block diagram symbols for these terms. (A block diagram is a highly specific form of a flow diagram.) The block diagram completely specifies the program to be executed. From it, trained assistants can punch onto cards its equivalent in machine-readable code.

An important feature for the user of GPSS III is simplified specification of probability distributions. These can be used to govern the times transactions arrive and the lengths of time they use facilities or storages. The GPSS III program moves transactions through the system according to the pathways specified in the block diagram and according to the times generated by the probability distributions. Thus, GPSS III is a generalized software facility for all-computer Monte Carlo simulation of the class of problems known as waiting-line or queueing problems.

Another attractive feature of GPSS III is the easy capture of statistical information from a simulation run. Statistical reports for facilities at the end of a run include: (1) for a facility—average utilization, number of entries, average time of service; (2) for a storage—its capacity, average contents, average utilization, number of entries, average time spent in the storage, and its maximum and ending contents; and (3) for queues—maximum contents, average contents, total entries, frequency that the queue was empty, average time in the queue, and the average cost of waiting.

The example problem described above written as a GPSS III block diagram is shown in Figure 9.2. A portion of the computer-readable input is shown in Figure 9.3, and some of the printed output is shown in Figure 9.4.

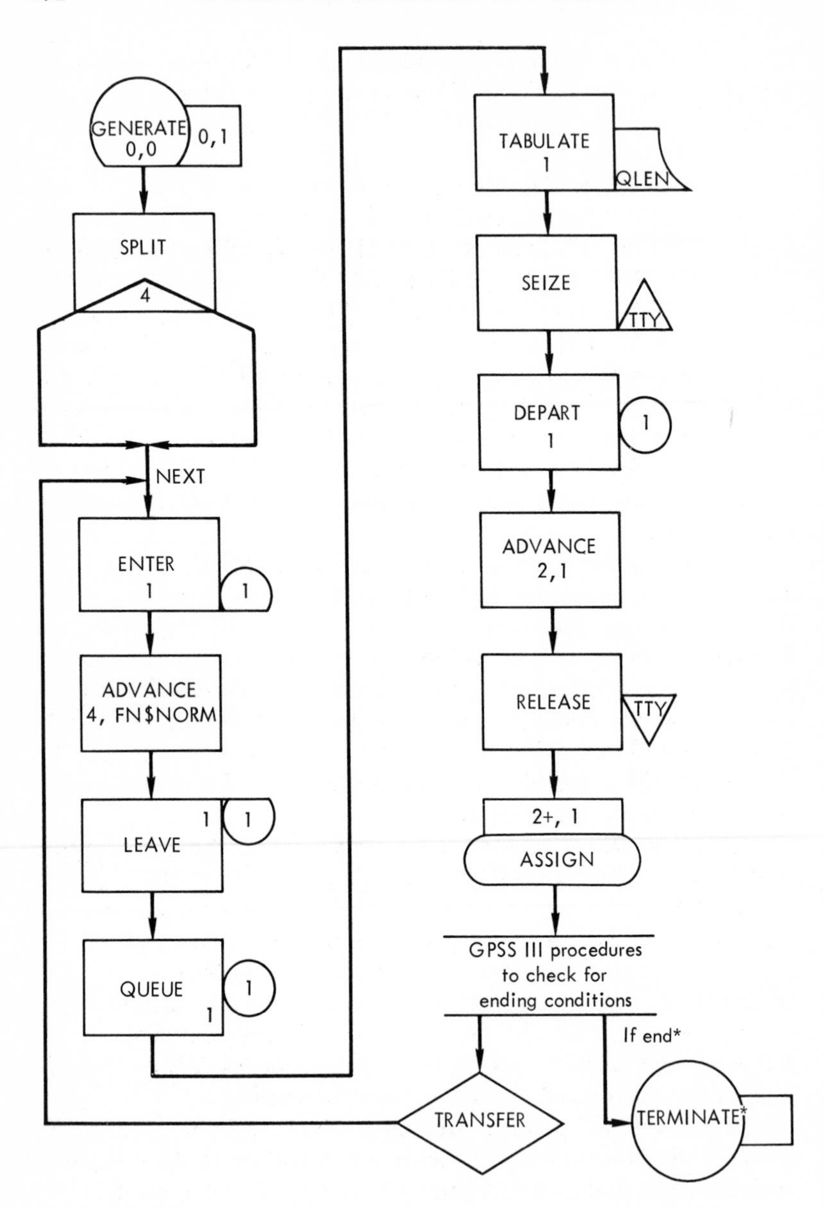

FIGURE 9.2. GPSS III Block Diagram for Example Problem

* If 50 bids for a pair, TERMINATE that pair; if open shop period is over, TERMINATE simulation.

```
*
* ALL CLOCK TIMES ARE IN MINUTES
*
* BID PREPARATION TIMES ARE NORMALLY DISTRIBUTED
*
      GENERATE   0,0,0,1     START AT TIME 0, GENERATE A PAIR
      SPLIT      4,NEXT,1    CREATE FOUR MORE PAIRS OF BIDDERS
 NEXT ENTER      1,1         PAIRS ENTER BIDDING AREA
      ADVANCE    4,FN$NORM   GENERATE BID PREPARATION TIME FOR PAIRS
      LEAVE      1,1         PAIRS LEAVE BIDDING AREA
      QUEUE      1,1         AND FORM QUEUE AT CONSOLE
      TABULATE   QLEN,1      KEEP A RECORD OF QUEUE LENGTH
      SEIZE      TTY         NEXT PAIR APPROACHES CONSOLE
      DEPART     1,1         BY LEAVING QUEUE
*
* SERVICE DISTRIBUTION IS UNIFORM WITH MEAN 2 MINUTES AND SPREAD +-1
*
      ADVANCE    2,1         GENERATE CONSOLE SERVICE TIME FOR A PAIR
      RELEASE    TTY         SERVICED PAIR LEAVES CONSOLE
      ASSIGN     2+,1        KEEP A RECORD OF TURNS BY PAIRS
```

FIGURE 9.3. Portion of GPSS III Program for Example Problem

FACILITY	AVERAGE UTILIZATION	NUMBER ENTRIES	AVERAGE TIME/TRAN
TTY	.996	141	2.127

STORAGE	CAPACITY	AVERAGE CONTENTS	AVERAGE UTILIZATION	ENTRIES	AVERAGE TIME/TRAN	MAXIMUM CONTENTS
1	5	1.584	.316	145	3.289	5

QUEUE	MAXIMUM CONTENTS	AVERAGE CONTENTS	TOTAL ENTRIES	AVERAGE TIME/TRANS	CURRENT CONTENTS
1	4	2.401	144	5.020	3

FIGURE 9.4. Portion of Output from GPSS III Program for Example Problem

Simscript II [13]

Originally, SIMSCRIPT was developed as a Level 5 language to simplify the work of writing simulation programs in FORTRAN. Programs written in original SIMSCRIPT were translated by the computer into Level 4 FORTRAN programs, which in turn were compiled into executable instructions. The next version of SIM-

[13] P. J. Kiviat, R. Villanueva, and H. M. Markowitz, *The SIMSCRIPT II Programming Language* (Englewood Cliffs, N.J.: Prentice-Hall, Inc., 1968).

SCRIPT, known as SIMSCRIPT I.5, introduced language improvements and implemented SIMSCRIPT as a Level 4 language with its own compiler, thereby eliminating its dependence upon FORTRAN.

SIMSCRIPT II is a major revision of this language that incorporates new conveniences for the programmer, recent innovations in computer software, and, like SIMSCRIPT I.5, its own Level 4 compiler. SIMSCRIPT II retains the original world view.

The simulation programmer who chooses a dialogue of SIMSCRIPT, by virtue of this choice, also chooses to look at his object system in terms of entities which are identified and described by their attributes, and which are organized in various ways into sets. The state of an entity at any time is the specific values assigned to its attributes. Entities which have the same attributes, but not necessarily the same values for attributes, form a particular set of entities which can have a name. Examples are ships on the sea, called SHIP, with attributes TONNAGE, SPEED, LOCATION; families within a theater, named VIEWERGROUP, with attributes NO.IN.GROUP and NO.ADULTS; or the industries in our example problem, named PAIR, with attributes BID.NO, NEXT.BID.-READY.TIME, NEXT.LEAVE.CONSOLE.TIME, and statistical attributes such as CUMULATIVE.WAITING.TIME, CUMULATIVE.CONSOLE.TIME, and CUMULATIVE.BIDDING.TIME. This latter set of pairs would be established in a SIMSCRIPT II program by the statement, "EVERY PAIR HAS A BID.NO, NEXT.BID.READY.TIME, NEXT.LEAVE.CONSOLE.TIME, CUMULATIVE.WAITING.TIME, COMULATIVE.CONSOLE.TIME, AND CUMULATIVE.BIDDING.TIME."

In SIMSCRIPT II, a set is comprised of members. Set members may be entities or may be other sets. Entities and sets may "own" other entities or sets. In our example problem, every pair may own sets of individual entities representing the respective bidding, waiting, and console time for each job opportunity number. Elsewhere in this book, we have called these sets cycle-by-cycle time paths. Computer storage for these would be reserved in SIMSCRIPT II by the statement, "EVERY PAIR OWNS A BIDDING.TIME.-PATH, WAITING.TIME.PATH, AND CONSOLE.TIME.-PATH."

The relationships among sets and their members are accomplished internally in the computer by a system of pointers that relates sets to members, members to sets, and members to members. The sets of sets of entities of SIMSCRIPT II can also be thought of as lists of lists of items. As in IPL-V, the system of pointers is necessary because members of sets are not necessarily stored sequentially in computer memory.

Entities may be permanent or temporary. Membership in a set by an entity (or by another set) may be optional. Temporary entities may be created and destroyed. The values of entity attributes may be changing all during a simulation run. During the execution of a program, entities are filed in and removed from sets according to priority rules. In the example problem, the queue of pairs before the computer console would be a set with a first-in first-out priority rule.

Simulated activity occurs in a SIMSCRIPT II program by recording only the beginning and ending of the activity. Such points in time, either the beginning or the ending of an activity, are called events. Activities change the state of the simulated system, but the state changes are recorded only when events occur (i.e., either when the activity begins or when it ends). Like GPSS III, SIMSCRIPT II contains a built-in next-event timing routine. This timing routine is the master controller of a simulation run, and simulation terminates when no further events are scheduled. It can simulate weekdays, hours, and minutes directly, eliminating the need for users to translate from dimensionless to meaningful units of time.

Events are of two types: (1) internal events, called endogenous events in earlier SIMSCRIPT dialogues, and (2) external events, formerly called exogenous events. As defined in this book, external events are input variables, and internal events are generated variables.

Because instructions are executed only sequentially (not simultaneously), a problem occurs in digital computer simulation when two events are scheduled for the same moment in time. This problem was not discussed in the section on next-event simulation in Chapter 8. There, by not mentioning the problem, it was implied that ties for what was to happen next were broken arbitrarily. However, SIMSCRIPT II provides for breaking next-event ties by the values of one or more attributes.

Built into SIMSCRIPT II are parallel random number generators that can provide different lists of pseudo-random numbers in the same run for separate Monte Carlo segments. Also built in are numerous probability distribution functions that can generate random variates directly on call. SIMSCRIPT II also computes and displays the pertinent statistics of a simulation run.

To review, the user of a SIMSCRIPT dialogue views his object system as sets of temporary or permanent entities, each with attrib-

```
PREAMBLE
THE SYSTEM OWNS A CONSOLE, A QUEUE, AND A LIST.OF.ARRIVALS
DEFINE LIST.OF.ARRIVALS AS A SET RANKED BY ARRIVAL.TIME
PERMANENT ENTITIES
     EVERY PAIR HAS AN ARRIVAL.TIME AND A NUMBER.OF.BIDS.COMPLETED AND
     MAY BELONG TO THE LIST.OF.ARRIVALS, THE CONSOLE, AND THE QUEUE
DEFINE NUMBER.OF.BIDS.COMPLETED AS AN INTEGER VARIABLE
DEFINE TIME.OF.NEXT.DEPARTURE AS A ''GLOBAL'' VARIABLE
EVENT NOTICES INCLUDE ARRIVAL AND DEPARTURE . . . END

          MAIN
          READ N.PAIR, MAXIMUM.NUMBER.OF.BIDS, OPEN.SHOP.PERIOD
          CREATE EVERY PAIR . . .
          FOR EACH PAIR . . .
               LET BID.PREPARATION.TIME = NORMAL.F(3.0, .75, 1)
               LET ARRIVAL.TIME(PAIR) = BID.PREPARATION.TIME
               FILE PAIR IN LIST.OF.ARRIVALS          LOOP
          ''EARLIEST.ARRIVAL.TIME IS FIRST IN LIST.OF.ARRIVALS''
'BACK'    LET EARLIEST.ARRIVAL.TIME = ARRIVAL.TIME(F.LIST.OF.ARRIVALS)
          IF EARLIEST.ARRIVAL.TIME < TIME.OF.NEXT.DEPARTURE GO TO DEPART
          ELSE SCHEDULE AN ARRIVAL AT EARLIEST.ARRIVAL.TIME ''AND'' GO CHECK
'DEPART'  SCHEDULE A DEPARTURE AT TIME.OF.NEXT.DEPARTURE
          ''CHECK TO SEE IF OPEN.SHOP.PERIOD IS OVER''
'CHECK'   IF TIME.V > OPEN.SHOP.PERIOD GO TO OUTPUT ELSE
          ''IF THE MAXIMUM NUMBER OF BIDS HAS BEEN MADE BY EVERY PAIR,
          ''STOP SIMULATION OTHERWISE CHECK FOR NEXT EVENT BY GOING TO BACK''
          . . .
          START SIMULATION ''PRINT RESULTS . . .'' STOP END

          EVENT ARRIVAL
          IF QUEUE IS EMPTY GO TO TTY
'THERE'   FILE F.LIST.OF.ARRIVALS IN QUEUE GO CHECK.ARRIVALS
'TTY'     IF CONSOLE IS NOT EMPTY GO THERE
          FILE F.LIST.OF.ARRIVALS IN CONSOLE
          LET SERVICE.TIME = UNIFORM.F(2.0, 4.0, 1)
          LET TIME.OF.NEXT.DEPARTURE = TIME.V + SERVICE.TIME
'CHECK ARRIVALS'  IF LIST.OF.ARRIVALS IS NOT EMPTY GO AHEAD
          SCHEDULE A DEPARTURE AT TIME.OF.NEXT.DEPARTURE
'AHEAD'   RETURN END

          EVENT DEPARTURE . . .
          FILE F.CONSOLE IN LIST.OF.ARRIVALS
          REMOVE THE PAIR FROM THE CONSOLE
          IF THE QUEUE IS EMPTY GO TO ALL
          FILE F.QUEUE IN CONSOLE
          REMOVE THE FIRST PAIR FROM THE QUEUE
          LET SERVICE.TIME = UNIFORM.F(2.0, 4.0, 1)
          LET TIME.OF.NEXT.DEPARTURE = TIME.V + SERVICE TIME
'ALL'     RETURN END
```

FIGURE 9.5. Portion of SIMSCRIPT II Program for Example Problem

utes that have individual values. As time goes forward, events occur that change the state of the system by either creating or destroying entities, or by changing their attribute values. These events may arise internally to the user's model, or they may originate externally. The simulation stops when there are no further events to be executed.

Portions of the example problem written in SIMSCRIPT II are shown in Figure 9.5.

Dynamo [14]

DYNAMO is a special-purpose compiler for the simulation of information-feedback systems. It was developed to serve a method of systems analysis for management called Industrial Dynamics.[15] While Industrial Dynamics is an approach to industrial management problems, the DYNAMO compiler may be used for other kinds of problems for which an information-feedback world view is appropriate. The conventions of Industrial Dynamics and the rules of DYNAMO together form a computer simulation language that for convenience we shall call simply DYNAMO.

The world view of DYNAMO sees the object system as a set of parts with interdependent interactions between them. A special flow diagram language for DYNAMO models of managerial problems has been developed. Some of its symbols are shown in Figure 9.6.

A DYNAMO model is a set of reservoirs, called "levels," interconnected by flows. The flows are governed by decisions that control the rates of flow. Inputs from outside the model come from sources, and outputs from the model go to sinks. Information can flow from any "level" of the model or from sources. Auxiliary variables collect information on which decisions depend. Where this information is about the consequences of past decisions, a feedback control occurs within the model. As time progresses, the particular structure of feedback control causes changes in rates of flow and consequently in levels. Decision rules compute the rates of flows on the basis of in-

[14] Alexander L. Pugh, III, *DYNAMO User's Manual* (Cambridge: M.I.T. Press, 1961).

[15] Jay W. Forrester, *Industrial Dynamics* (Cambridge: M.I.T. Press; New York: John Wiley & Sons, 1961.

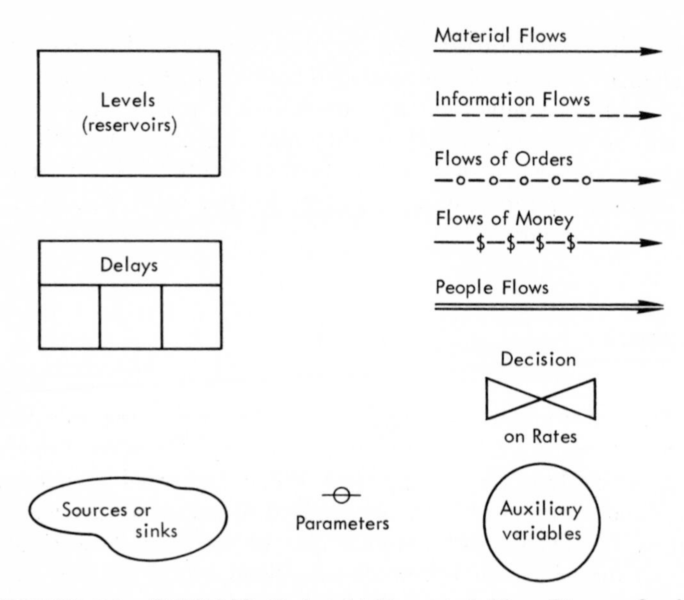

FIGURE 9.6. DYNAMO (Industrial Dynamics) Flow Diagram Symbols

formation flowing to the decision rules. The set of decision rules in a model constitutes a simulated overall policy for governing the simulated object system.

The output of a DYNAMO simulation run is usually a graph showing the values for various rates, levels, and statistical variables over the entire simulated time period. Through these graphs, simulation runs can develop the time-dependent implications of alternative system structures and policies.

Industrial Dynamics originally viewed object systems very mechanically. The possibility of adaptive behavior by the decision makers that would change the decision rules was not allowed. Subsequently, DYNAMO was implemented on time-sharing computing systems, and some attempts to run adaptive man-computer simulations with live persons have been made. Also, attempts to incorporate intangibles are underway.[16]

Our GPSS III and SIMSCRIPT II models of the example problem used next-event timing routines. In fact, the flow diagram of Figure 9.1 is a next-event flow diagram. DYNAMO, in contrast,

[16] Edward B. Roberts, "New Directions in Industrial Dynamics," *Industrial Management Review*, Vol. 6, No. 1 (Fall, 1964), pp. 5–14.

operates with a fixed time increment cycling mechanism. The length of this increment is labeled DT, and it may stand for a number of simulated minutes, hours, days, weeks, months, or years. DYNAMO looks behind one DT to a past moment in time called J, and ahead one DT to a future moment in time called L. The present is called K. In operation, DYNAMO computes at time K, the new levels for K, using the rates of flow for the time period J to K, labeled JK. Then, from information about the new levels, it computes the new rates of flow for the time period from K to L, labeled KL. For the next cycle, DYNAMO shifts time K, the present, to time J, the past; shifts time L, the future, to time K; moves the rates of flow for period KL to period JK; computes new levels for the new time K and new rates of flow for the new period KL; shifts time forward one DT (K to J, L to K, KL to JK); computes, shifts time, and so on until the desired total time is simulated.

Let us think of the example problem queue, computer console, and number of pairs preparing their bids as levels (or reservoirs). Of course, the console as a reservoir can have only levels zero or one pair, and the level of the queue or of the pairs bidding can never exceed the total number of pairs simulated. See the boxes in Figure 9.7.

In DYNAMO, the choice of the length of DT is very important. In the case of the example problem, DT must be long enough so we can sensibly model a flow of pairs from the console. Say we choose DT so that the average number of pairs from the console is five. This means five simulated pairs on the average will leave the queue, approach the console, complete their interaction with the computer, and leave the console. Now, think of pairs entering the console area and departing at rates different from the average. In these cases, they are either delayed or accelerated compared to the average flow through the console area. Let's call this view of pairs using the console a "delay."

In DYNAMO, a delay is a special combination of a level, a decision rule, a delay parameter, and a DT or fraction thereof. Hence, the simulated human variability of pairs using the computer console is represented by delays (or accelerations) of the flow of pairs from the console during a DT. DYNAMO creates this effect through auxiliary random influences on decisions called "noise." For our example problem, random influences are needed at the console delay and for the flow of pairs to the queue.

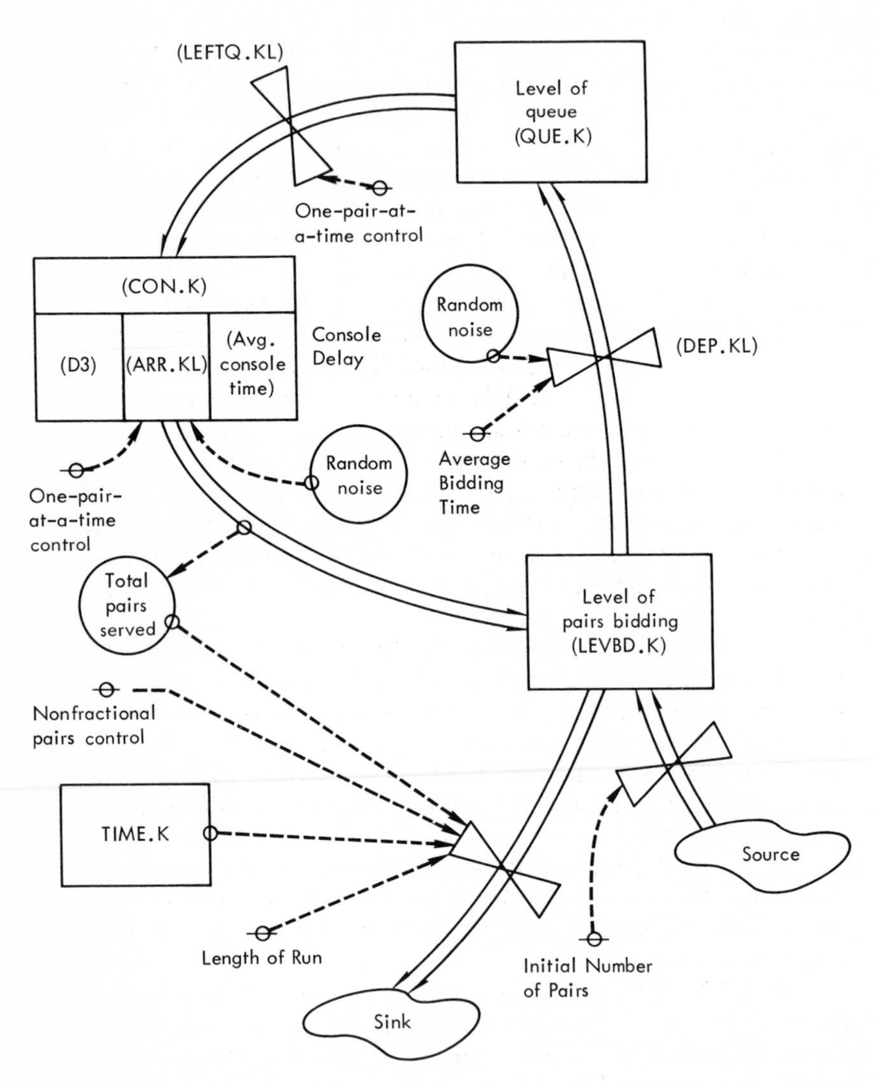

FIGURE 9.7. DYNAMO (Industrial Dynamics) Flow Diagram of Example Problem

Finally, let us specify the decision rules shown in Figure 9.7 as follows:

1. The DYNAMO delay which governs the flow away from the console will depend on the average number using the console in a DT and on random noise.
2. The rate of flow from pairs bidding will depend on the level of pairs bidding, the flow from the console, and on random noise.
3. The rate of flow from the queue will depend on the console delay.
4. The initial queue from the external source is given by a parameter.
5. The flow to the external sink will depend on the current clock time, the ending time for the simulation run, and the number of pairs served by the console.

A segment of a DYNAMO program for this model is shown in Figure 9.8.

```
NOTE  LEVEL OF PAIRS BIDDING
1L    LEVBD.K=LEVBD.J+(DT)(ARR.JK-DEP.JK)
NOTE  NOISE (RANDOM VARIATIONS IN BIDDING TIMES)
34A   VAR.K=(1)NORMRN(0,.2)
NOTE  DEPARTURE RATE FROM TIME K TO TIME L
39R   OUT.KL=DELAY3(ARR.JK,9)
NOTE  DEPARTURE RATE WITH NOISE
7A    AUX.K=OUT.JK+VAR.K
7R    DEP.KL=CLIP(LEVBD.K,AUX.K,AUX.K,LEVBD.K)
NOTE  CHECK TO SEE IF QUEUE IS EMPTY
51A   AVAIL.K=CLIP(1,0,QUE.K,1)
NOTE  GO TO CONSOLE IF A PAIR IS IN QUEUE AND CONSOLE VACANT
51R   LEFTQ.KL=CLIP(0,AVAIL.K,CON.K,1)
NOTE  QUEUE LEVEL AT TIME K
1L    QUE.K=QUE.J+(DT)(DEP.JK-LEFTQ.JK)
NOTE  CONSOLE LEVEL AT TIME K
1L    CON.K=CON.J+(DT)(LEFTQ.JK-ARR.JK)
NOTE  NOISE (RANDOM VARIATION IN CONSOLE SERVICE TIMES)
34A   NOISE.K=(1)NORMRN(0,.5)
NOTE  ARRIVAL RATE AT BIDDING AREA WITHOUT NOISE
39R   GO.KL=DELAY3(LEFTQ.JK,3)
NOTE  ARRIVAL RATE AT BIDDING AREA WITH NOISE
7A    BID.K=GO.JK+NOISE.K
7R    ARR.KL=CLIP(1,0,BID.K,1)
NOTE
NOTE  INITIAL CONDITIONS
NOTE
6N    LEVBD=5
6N    QUE=0
6N    CON=0
NOTE
PLOT  LEVBD=L,QUE=Q,CON=C
SPEC  DT=1.0/LENGTH=300/PRTPER=5/PLTPER=1
```

FIGURE 9.8. Portion of DYNAMO Program for Example Problem

EXERCISES

9.1. This book asserts that a model is a specific expression of a theory. Therefore, a computer model is also an expression of a theory. Discuss the implications for expressing theories of constructing models in computer languages with different world views.

9.2. Design a language for man-computer conversations to obtain responses from live subjects. Assume you already know a Level 4 language; hence, this assignment is a Level 5 language problem.

9.3. Using the verbal commands for GPSS III shown on page 191, draw a flow diagram for a model of an object system that is a typical barbershop. Write the name of the command in capital letters in the boxes of your flow diagram, and any descriptive additional words in upper- and lower-case.

9.4. Think of an object system from your experience that might be interesting to simulate. For this system, define entities and sets of entities in the manner of SIMSCRIPT II. Be sure to list all pertinent attributes of entities.

9.5. Using the special flow diagram symbols for DYNAMO and Industrial Dynamics shown on page 198, design a simulation of an object system from your experience for which the world view of DYNAMO is appropriate.

9.6. At a bazaar to raise funds for charitable purposes, a masked citizen sits in a booth and offers to match coins with any person. He claims to be a master of his art. The bazaar is crowded. There are many other booths. The masked master coin flipper allows his clients to choose any denomination coin and to match with him any number of times. Discuss the suitability of GPSS III, SIMSCRIPT II, and DYNAMO for modeling this object system.

9.7. Prepare in flow diagram form (using the commands from page 191) a model structured with the GPSS III world view for the bazaar booth system of Exercise 9.6. Write the name of the command in capital letters in the boxes of your flow diagram, and any descriptive additional words in upper- and lower-case.

9.8. Define for a SIMSCRIPT II model of the bazaar booth

system of Exercise 9.6 the following: entities, attributes, sets, internal events, and external events.

9.9. Prepare in flow diagram form (using the symbols from page 198) a model structured with the DYNAMO world view for the bazaar booth system of Exercise 9.6.

10

A SURVEY OF SIMULATION APPLICATIONS

It is difficult to find object systems that have not been modeled or simulated in one way or another. However, only since the advent of electronic computers, have amounts as large as a million dollars been expended on single simulation projects.

Some simulation projects stop before they are completed. Others continue, and their supporters and workers speak in terms of hope. Some are outstandingly successful.

Except in education, successful simulations are seldom publicized. A security barrier prevents publication of many military simulations. In private industry, proprietary interests prevent public knowledge of successful simulations. Yet the multitude of object systems that have been simulated and publicly described is so vast that a survey with complete bibliographic documentation would require several books. Moreover, different observers of the simulation milieu would develop surveys that would vary according to their individual interests and professional experiences.

Offered in this chapter is a summary of the variety of simulations the author has learned of over a period of more than five years. These projects are reported in broad categories according to a taxonomy that encompasses both general types of object systems and general purposes for conducting the simulations. No attempt is made to provide references or to describe each simulation in detail. Occasionally, a project will be identified as a man-model, man-computer, or all-computer simulation. Many of the projects mentioned are pilot studies or efforts to establish simulation as a method for the subject area mentioned.

A selected bibliography of general references appears at the end of this book. Included in this bibliography are other bibliographies and reports of symposia and conferences. Readers can find descriptions of many more applications in these references. Also, simulation is sufficiently established that reports of applications appear regularly in technical and professional journals and occasionally in the popular press. Relevant journals are also listed in the bibliography. The organizations sponsoring these journals hold regular meetings at which simulations are described and simulation methodology is discussed. The published programs and proceedings of these meetings are an excellent source of reports of recent simulation work.

Simulation of Computer Systems

Computers are efficient means for conducting simulations. They are also object systems frequently simulated by other computers. Of course, a computer simulating a computer is not simultaneously simulating itself, although it may be simulating itself at another or hypothetical time. Existing computers, computer systems, and simulation techniques may be used to study hypothetical alternative future object systems that are computers, computer systems, and simulation techniques.

All-computer simulations have been used to study high speed computer logic, complete computer systems, the performance of such systems under various conditions of usage, computer networks, alternative computer designs, particular types of computer services such as time sharing, and even a complete computer center including its users.

Paradoxically, complex computers are sometimes used to simulate simpler computers. As new computers are acquired, they usually have different instruction sets. Manufacturers frequently supply simulation packages that enable computer purchasers to run old programs immediately. Thus, until new programs are written, the new computer simulates the old computer to run the old programs.

Implementation of a high level language such as FORTRAN on a computer system begins with machine language and proceeds through the development of assemblers and compilers. Interestingly, this complex chain of interrelated computer software systems has been used to simulate the elementary computer logic on which it is all based. The purpose of this "reverse" procedure is to study the effect of changes in the basic instruction set.

Computers are used to control industrial processes such as oil refining and material handling systems such as automatic warehouses. Simulations of these computer-controlled systems, developed during design stages, are in effect simulations of computers by other computers.

Educational Gaming

Games that simulate subject matter to be learned are increasing in use. While some educational games are complex, many are structured around only a few concepts from the subject area. Their use ranges from elementary grades through undergraduate and graduate college years to continuing education programs such as management development seminars. Their complexity spans single players (or teams acting as single players) interacting with simple models to coordinative players organized as teams interacting with intricate model segments, and these teams then interacting competitively with other teams. A frequently stated purpose of interactive coordinative-competitive games is to enable participants to understand better the complex and subtle processes of behavior in organizational settings. This is particularly true of games designed for training in educational administration. Simulation materials have been developed for the roles of school principals, school superintendents, and for various assistants.

The discipline of economics presents examples of games to

teach basic concepts. Separate games are commercially available for teaching concepts of the market, the firm, collective bargaining, the community, scarcity in allocation, banking, the national economy, and international trade. In political science, educational games are available for simulation of international relations, elections, a legislature, and disasters. Games for elementary and secondary schools have been devised for diverse object systems such as caribou and seal hunting, career selection, parent-child relationships, neighborhood planning, pollution problems, preparation for reading readiness, the industrial revolution, politics, crime problems, and foreign policy.

Other educational games deal with internal revolutions, strategy shaping, creative mathematics, propaganda, catastrophes, crises, maintaining social order, and accounting and inventory systems. One inter-nation simulation requires decisions by five to seven nations, each of which has specialized decision makers for internal affairs, external affairs, and military forces and a central decision maker whose position is threatened by an aspiring out-of-office central decision maker. Some man-model educational games, such as the one simulating a legislature, have lead to all-computer simulations in efforts to understand better the behavior of the educational participants. Others—one about ruling a city-state complex in 3500 B.C., two simulating management of a toy store and a company manufacturing surfboards, and one where players are advisors to young nations—have been implemented on dedicated time-sharing computers for sixth-graders who sit at remote terminals. Some computer-based educational games use Monte Carlo simulations in the computer model to illustrate the uncertainties of life.

Elsewhere, we noted that computers can simulate computers. For educational purposes, console mock-ups of new computers have been connected to old computers to train operators for the new computer.

Simulated Counseling

Counsel can be obtained or given on about anything. Attempts to simulate counseling activities usually involve man-computer interaction, with the computer taking the role of the counselor. One major project is attempting to provide computer implemented vocational counseling and guidance. Another project substitutes the computer for a school counselor to give advice on courses.

Educational Administration

The problem of matching the needs and constraints of students, teachers, and available space has long plagued school and college administrators. Computer-based simulation approaches to this problem attempt to reduce the phenomenally large number of combinations possible in academic scheduling to a few feasible alternatives that can be evaluated by academic administrators.

Academic scheduling by computer begins with the classes that are to be offered, the instructors and rooms available, the needs of students, and the times at which classes may be offered. The program then assigns times, instructors, rooms, and students to classes to simulate one alternative schedule. Usually the first result is not satisfactory. Administrators then adjust their ideas of what a master schedule should be like, and the progam is run again. This procedure is repeated until a satisfactory schedule is finalized.

Once an academic scheduling simulation was used to help design a community college. By running and rerunning the simulation with different assumptions about rooms and room sizes, decisions were made that reduced the number of rooms to be constructed by more than twenty percent and doubled the utilization per room compared to similar colleges.

Another application simulates overall university operations under a number of different conditions in an effort to avoid costly mistakes before committing funds. This model integrates information on staff, space, students, finances, and administrative decisions for a single institution. For school districts, one major project has represented staff, facilities, overhead, and teaching technology on a budget and output basis; another has represented the internal organization of schools to study problems of innovation in instructional systems.

Life and Health

All-computer Monte Carlo simulation is particularly suitable for representing dynamic life processes. Applications in this area include studies of marriage, fertility, natality, and mortality. In fact, entire life processes, including competitive evolution, have been simulated for such object systems as lobsters, whooping cranes, viruses, apple trees, animal genetics, epidemics, and even the evolution of forests under man-controlled timber management. Processes of mental health such as perceptual development and neurophysiological behavior have been represented by all-computer simulation. Computer programs have deen developed that intentionally behave neurotically (although programmers regularly believe computers to be neurotic in their own right).

Administrative decisions for health, hospital, and medical services are particularly suited for assistance from simulation. Examples are the simulation of alternative admissions policies, alternative means for training doctors, and the scheduling of a multiple operating room system. The operation of a maternity ward and of a psychiatric out-patient clinic have been simulated. Community health planning has been represented by all-computer simulation. Man-model simulations of the effects on a community of an epidemic and of the community's attempts to detect, halt, or ameliorate these effects have aided planning for such emergencies. Finally, the diagnostic decision processes of physicians have been studied by displaying computer generated symptoms on time-sharing terminals to actual physicians who then responded with diagnoses.

Other life object systems simulated by computer are plasmas, sleep, lung functions, muscle actions, blood circulation, and radioisotope localization in the body.

Man-Machine Systems

Most man-machine simulations involve some nondigital physical analog of the object system. Examples are instrument flight trainers and indoor golf courses with motion picture simulations of fairways. However, for complex man-machine systems, digital computers are finding applications. In these cases, digital computers are connected to the analog simulation device through special converters. An example is a lifelike dummy used to train anesthesiologists. The dummy simulates a surgical patient and reacts to anesthetics through a digital computer program that represents human bodily responses.

Experimental Games

Game theory is a system of analysis that assumes certain rules of player behavior. Based on these assumptions, ultimate or equilibrium consequences for games played by two or more players are deduced. In two-person zero-sum games, the amount won by one player is lost by the other. In nonzero sum games, the payoffs to each player for each combination of decisions are specified. In the 1940's and 1950's, an elegant analytical theory was developed that so tantalized behavioral researchers they undertook experiments with live subjects to see whether the assumptions and conclusions of game theory were valid. This was done while overlooking the purport of original game theorists that their assumptions were not about live human beings, but merely a point of analytical departure. Experimental games now flourish to investigate variables far beyond any considered by analysis in formal game theory.

The game probably used more than any other for live behavioral experiments is the Prisoners' Dilemma. The general form of this model specifies two uncommunicating players who each have two possible decisions that interact to determine the individual payoff to each. The game is epitomized by the story of two prisoners—call them A and B—charged with a joint crime and held incommunicado. If Prisoner A confesses and Prisoner B does not, Prisoner A is promised a three-year sentence and Prisoner B will receive nine years, and *vice versa*. If both confess, each will receive six years. If both hold out and do not confess, each will receive only one year. Neither can control or even know what the other will do. If A holds out, B is better off to confess and receive three years rather than six years, and *vice versa*. If A confesses, B is better off to confess and receive six years rather than nine years, and *vice versa*. Hence, not knowing what the other will do, each is better off by confessing. Yet, if both held out, each would receive only one-year sentences. This situation is not unlike gasoline price wars, competition among automobile manufacturers, or competitive bidding for contracts which are examples of market structures economists call *oligopolies*. Oligopoly itself has been investigated by many man-model simulations.

The numbers used above to illustrate the Prisoners' Dilemma are only a few that can be chosen that retain the desired relationships. In an experiment, the payoffs may represent real or fictional rewards. Holding out by one prisoner implies trust of the other. Confession by one implies distrust of the other. Holding out is cooperative behavior; confessing is conflict behavior. Coordination is relevant when some degree of communication or repetitiveness is permitted. The appeal of Prisoners' Dilemma games is now apparent. Here is a man-model simulation vehicle in which variations of payoffs, trust, cooperation, conflict, communication, repetition, and coordination can be studied.

When degrees of communication and agreement are permitted, models similar to the Prisoners' Dilemma simulate bargaining behavior. One research team has implemented bargaining and negotiation experiments on a dedicated computer that controls up to 24 remote terminals. Experimental variations of Prisoners' Dilemma, oligopoly, bidding, and bargaining models have been conducted to study the influences of winning and losing, differences in motives and levels of aspiration, need to maintain face, different presentation formats, knowledge of the other player or of his strategy, side payments, personality and attitude attributes, emotional relationships among players, threats, sex, race, prenegotiation behavior, risk-taking propensity, pacifist strategies, compromise and other means of conflict resolution, boredom, family background, interference by "outsiders," and outright attempts to induce collaboration.

Simulation of Human Thought

Human thought self-destructs the moment it is reflected upon. Hence, introspection invalidates itself as a research method by loss of its object of research. Outsiders have yet to gain access to another's mind. The loneliness of individual thought remains unpenetrated. One denial that computers can think is the very fact that their processes do not self-destruct but can be replicated for study and can be accessed by outsiders—this is in distinct contrast to human thought. However, this fact of replication of and access to computer processes has brought forth a methodology for studying human thought, in particular the cognitive and problem solving processes.

The operation of a computer program can be studied directly, while the operation of the human mind cannot. It has been argued that if a computer can be programmed to behave like a human, the program that does the trick is itself a valid theory of the human behavior simulated. Studying the dynamic behavior of the program—which can be done—then lends insight into the dynamic processes of the human mind that produces the object system behavior. Stated another way, it is claimed that one must understand human behavior to simulate it by computer, and the computer program that simulates it successfully is an expression of that understanding and hence a theory of the human thought producing the behavior.

Of course, if one were successful in simulating human cognitive or problem solving behavior by mathematical equations, the implication is that humans think in mathematical equations, or at least their equivalents. The author has attempted to simulate live participants in a man-model competitive bidding simulation by as many analytical models as he could conceive—all to no avail. The conclusion is that analysis fails to capture the nuances of dynamic human decision processes. Other workers, using rules of thumb called heuristics, have quite successfully represented the decision behavior of subjects in binary choice experiments, subjects proving simple theorems, the decisions of an investment trust officer, verbal learning, and certain search behavior.

All-computer simulations have also represented the internal mental processes of concept attainment, belief systems, pattern recognition, puzzle solving, trial and error groping toward deductive reasoning, short-term memory, synonym generation, risk taking, crisis reasoning, and reinforcement of behavior.

Many of these efforts begin with subjects' "protocols"—records of their "thinking aloud" explanations of their own thought steps. Protocols attempt to probe behind human decisions at the moment the decision becomes overt behavior. Man-computer simulations and online administration of behavioral experiments provide an exceptional opportunity to use the computer as both interrogator and as record keeper for protocol creation.

Communication Networks

An early well-known man-model simulation represented the effect of alternative communication networks on group problem solving effectiveness. Subsequently, all-computer simulations of the original man-model experiments were developed. These efforts deal with structured interaction among small groups of human decision makers. Another all-computer simulation investigated the organizational effects on a larger structured interacting group—a wholesale lumber market—of varying communication channels and costs.

Other simulations of communication networks deal with massive systems such as telephone networks. These models are seldom concerned with the content or effect of the messages, merely their volume. There are even digital simulations of human speech to represent the transmission of sounds in a room, and of pictorial displays to represent the effects of "noise in the picture." One study dealt with the effect of emergencies on communication networks, including the behavior of human telephone operators; another represented the diffusion of information in a region.

Organizations

In addition to the communication aspects of large and small groups mentioned above, man-model and man-computer simulations have been developed to study organizational aspects of policy implementation, acceptance of recommendations and changes, bureaucratic authority, learning, leadership, innovation, latent motives of members, biases in decision making, and interpersonal and face-to-face behavior.

In one man-model organizational simulation, subjects using tape, scissors, and cards simulated manufacturing tasks. Another man-computer simulation vehicle represents complex hierarchical organizations with live participants located at 24 terminals connected to a dedicated central computer. In this implementation, the participants communicate with each other only by messages received, sent, and recorded by the computer. Such simulations with live participants permit evaluation of the effectiveness of alternative organizational configurations.

Simulation of Social Systems

Social systems are not as structured as organizations and hence present challenging modeling problems. All-computer simulations of social interaction have traced the patterns of interpersonal contacts and rewards, conflicts in social roles, and social pressures. The rigors of computer modeling have revealed logical flaws in previously accepted sociological theory.

Other studies in this area, principally man-model simulations, have investigated conflict and aggression in social processes, kinship and its relation to social structure, collective behavior during panic, attitude convergence, social feedback and the need for social approval, control through social pressures, social mobility, and patterns of conformity and deviation.

Civil, Political, and Community Simulations

Object systems from the public domain studied by simulation include the processing of criminal cases through courts, legislative redistricting, voting paradoxes, referendum controversies, municipal budgeting, police and public safety procedures, public transportation networks, sewerage systems, the war on poverty, and community disasters such as floods or fires.

Rural and urban standards of living have been simulated, as well as problems of urban growth. Man-model simulations for training of urban planners have become popular.

Regional, National, and International Systems

All-computer simulation in the form of dynamic repetition of nonstochastic analytical models is particularly suited to representation of political and economic characteristics of regions, nations, and international systems. Aspects of political economies that have been simulated are development and trade problems, underdevelopment, national elections, the effects of changed tax laws, poverty, the effect of transportation alternatives on economic development, and the impact of cultural variations. Other projects have represented complete socio-economic systems and designed models for national planning that incorporate submodels of particular economic segments such as the residential housing market. DYNAMO and Industrial Dynamics provide a convenient vehicle for representing the information-feedback systems of a national economy.

Man-model and man-computer simulations of developing economies, international peace and war, and internationalism and isolationism are used for both educational and research purposes. This includes the international simulation mentioned under Educational Gaming.

Object systems that are broad markets have been frequently represented. Examples are the United States plywood industry and the home mortgage market.

War and Crisis Gaming

Man-model war gaming as it is practiced today evolved over the centuries from entertainment games like checkers and chess, which are played on boards containing 64 squares. Many war games are board or map games in which various kinds of tokens are moved about to simulate military maneuvers. One version, called *Kriegsspiel* (literally "war game"), 3600 squares and a 60-page rule book. A subsequent version, called *Free Kriegsspiel* emphasized control by an umpire instead of by rigid rules. During World War II, war games and map exercises aided in making actual military decisions. One instance is reported in which an attack occurred while a map rehearsal game was underway. The game was ordered continued, but actual combat reports were substituted for simulated data.

Man-model war and crisis gaming is called three-room simulation when two opposing teams are segregated in meeting rooms and a control or umpire team is situated in a third room. The teams in conflict interact with each other only through the control team. The initial model for this type of simulation is given to the participants in the form of a scenario and perhaps a fact book. Sometimes additional scenarios are given during the course of a game run. Much of the model is developed by the control team as the simulation proceeds.

When the simulation deals with military strategy and tactics in combat situations, these techniques are known as war gaming. When the simulation deals with political-military or cold war conflicts, they are called crisis gaming.

Operational Gaming

Alternative management and administrative structures for large logistic systems have been explored in special laboratories, particularly the Logistic Simulation Laboratory of the Rand Corporation in Santa Monica, California. These exercises or experiments involve large numbers of live subjects with both computer and noncomputer models. The subjects are either specially trained for their roles or are recruited from the object system itself. While the past significant developments in operational gaming have come from massive and lengthy projects conducted for military and defense purposes, operational gaming may eventually be set free of its present restriction to particular laboratories and full-time subjects. The futuristic executive decision room described later under Simulation Aids to Business Decision-Making combined with the remote access features of the dream management game mentioned below suggest new forms for operational gaming.

Management Games

Following the pattern set by war gaming, but capitalizing on the development of electronic computers, management associations and university business educators have developed man-computer simulation models of business firms. These vary in complexity and are used for indoctrination, specific training, or general education. They cycle on fixed time increments that represent months, quarters, or years. Participants, either as single players or as members of simulated management teams, make numerical decisions for variables such as product prices, production levels, sales efforts, and materials purchases. Their decisions are then combined with carryover data from the last simulated period to become input for the next cycle. The output from the model represents the interactions among and within business firms and forms the basis for the next round of decisions. A simulation run stops after a given number of cycles, which may include a final set of cycles with "frozen" decisions to keep players from sacrificing the future for the sake of winning the game in the present.

Management games have been written for many specific industries and products. Usually one specialized function of business such as marketing or production is emphasized. Some are structured to illustrate or provide practice in special skills and concepts, e.g., the dynamics of accounting records. Some management games do not require computers but use playing boards and tokens. Some represent abstract products. A man-computer game developed by the author can simulate manufacturing industries for any desired product class merely by changing parameters. In this game, three different products in two classes may be manufactured by each of two to five competing firms. Product and labor market responsiveness can be adjusted to the industry being simulated. Also, qualitative judgments of referees or umpires may be included as input data to the computer model. This game can shift from years to quarters to months during a single run.

Management games are used mostly in college courses and in management development programs. Large numbers of business students and company employees have played them. A few attempts to use them as research vehicles have been made, but management game runs are usually too loosely structured and generate data for too many variables to produce conclusions meeting research criteria. However, it is just this interaction among variables that these games attempt to teach participants.

In some applications, management games are played in single sessions with decisions made every twenty minutes and results of each decision returned in ten minutes from a batch processing computer. In others, decisions are made once a week with results returned at the next class period. An unrealized dream of management gamesters pictures a game model running periodically at a central computer in some ratio of simulated time to real time. From remote terminals from anywhere in the world, players can inquire the status of their simulated departments and companies. They can then either change decisions or let their old decisions continue. The computer would inexorably execute the game model at each appointed time using the decision values existing at that moment. This dream also includes the availability of the general game model in a time sharing mode to pretest decisions before finalizing them for insertion into the "real" simulation run.

Queuing Systems

A major application of Monte Carlo simulation is representation of traffic problems. The simulation language GPSS is particularly suited for this purpose because it deals with "transactions" flowing through a system of "facilities" and "storages." Simulation has been used to study vehicular traffic, transportation policies, message traffic in telephone networks, radio-dispatched fleets, shipping schedules, scheduling jobs through shops, taxi fleets; air transportation systems, movement of replacement parts, cargo handling, dock management, aircraft movements on the ground, distribution networks for products, jobs through a computer center, the routing of railroad cars, and the example problem of Chapter 9.

Most of these problems are characterized by waiting-lines or queues. This class of problems presents a classic case of the failure of mathematical analysis to represent real object systems. There is an elegant mathematical theory of queuing systems, but it has not yet been extended to situations of reasonable complexity such as those involving multiple, parallel, or sequential channels through which items of traffic must move or wait for service. In the general queuing problem, either arrivals are waiting and the system is busy, or there are no items being serviced and the system is idle. This situation may occur at many intermediate points in a total system. The managerial problem is to achieve some balance between costs of idleness and the costs of waiting. The reason that idleness cannot be perfectly matched to the available service is that complete managerial control is absent. Thus, most queuing systems are stochastic in character and, where analysis fails due to complexity, Monte Carlo simulation carries on.

Operations Research and Management Science

Queuing or waiting-line problems are one of several classes of problems that have come to be treated by workers known as operations researchers and management scientists. Some of the forms of problems in this general area are:

1. Queuing problems (discussed above): How to balance the costs of waiting against the costs of service.
2. Allocation problems: How to allocate resources among competing uses.
3. Inventory problems: What are the optimal sizes and ordering patterns for inventories?
4. Sequencing problems: How to arrange tasks in time.
5. The routing problem: Where and in what order to send persons or things?
6. The replacement problem: What is the optimal scheme for replacing parts or facilities that wear out?

Operations research and management science grew out of the applications of mathematics to administrative decision problems. It was soon discovered that many of these problems were too complex to be expressed in mathematical models that would yield analytical solutions. About this same time, general purpose digital computers were being developed and simulation in all its forms became a major tool of operations researchers and management scientists.

Simulation Aids to Business Decision Making

A dream of some management scientists pictures a corporate headquarters room called the game room or the simulation room. It is a futuristic chamber for the executive committee and the board of directors of the corporation. Time-sharing consoles are built into each chair. The executive in charge of decision simulation operates a master console controlling the individual terminals. There are large screens that show computer-generated outputs. Selected images also appear on individual screens at each executive chair. At the central computer, both all-computer and man-computer general models of the corporation, its environment, and its competitors are available online. Parameters and data to make the general models specific to any particular problem under discussion can be called into execution from any chair.

In this dream, executives explore decision alternatives by interacting with models, each supplying his own human decisions through his time-sharing console. They trace the consequences of hypothetical decisions in several modes. First, a general strategy is simulated for the present behavior of competitors and the present internal corporate organization. Second, behavior of competitors is hypothesized and tested against alternative decisions. Third, a sequence of interactive decisions is tested in which original decisions are modified in light of anticipated competitive decisions, which in turn alter competitors' behaviors, which in turn alter corporate decisions, and so on. In this mode, some members of the group may act as executives of a competitor by using parameters and data pertinent for that competitor as input to the generalized simulation programs. Finally, in any of the above three modes, the executives conceptually isolate themselves at their time-sharing consoles to simulate the corporate internal organization and its ability to adapt to environmental and competitive conditions.

The above dream has not happened. If it did, and the all-computer and man-computer simulation systems became effective means for making business decisions, then the simulations themselves could be considered as much a part of the corporation as are its executives. The implication is that the simulation and its object system become so intertwined that the distinction made in this book is lost. However, just as executive decisions are about future events, executive simulations likewise are about future events. Hence, the simulation is indeed about something else, which may in fact be the simulation itself and the corporation it is a part of, but at some future time under hypothetical conditions. The effect is an ongoing system of hindsight in advance, so to speak, built right into the information system of the corporation.

The dream executive decision environment described above contains simulation models of entire firms and their environments. Simulations that attempt to represent such comprehensive object systems usually become very abstract, and hence too general to be practical. Useful simulation aids to business decision making presently focus on smaller problems. Examples are simulations of budgets—cash, accounts receivable, credit policies, collections—and the classes of problems listed earlier under Operations Research and Management Science. In business applications, simulation has been used to design chemical plants and typewriters, to test stock market strategies, and to analyze new ventures under risky conditions. Simulations of problems of introducing a new product have considered the multidimensional attributes of alternating products, the timing of the introduction, and alternative advertising strategies.

Business-related all-computer simulations of human thought have represented the decisions of an investment trust officer, a department store buyer, wholesale salesmen, the reordering techniques of photographic dealers, the word-of-mouth flow of information in marketing channels, and consumer decisions on shopping. Man-model simulations have studied individual preferences for particular brands, field interviewing costs for market research, and collective bargaining strategies.

Other simulation aids to business decisions have represented agricultural soil testing in laboratories, the financial flows resulting from hiring and firing life insurance agents, alternative marketing channels and distribution systems, research and development strategies, a linen supply company, telephone repairing costs, and a grain elevator.

Personnel Problems

All-computer simulation has been used to study the personnel needs and manpower management problems of military branches where new men are required to fill vacancies and replace losses, and existing men are rotated among duty tours.

Man-machine, man-model, and man-computer simulations significantly contribute to the indoctrination and training of new personnel. One technique simulates the activities of a decision maker at his desk attending to mail in his "in" basket. He must act on each item. Other techniques involve simple man-model simulation with booklets that resemble school workbooks. Some of these include role playing and dynamic decision making. Simulation materials for training in supervisory skills, collective bargaining, production and inventory control, and purchasing have been developed.

For a public accounting firm, the author prepared a man-model simulation to orient new employees. The player role was a fictitious manager of a general office of this firm. Printed cards, pins, chips, and cork pads were used to implement the model.

Personnel managers sometimes want to use simulations to evaluate persons for selection or for promotion. When the simulated object system is a well-understood skill such as assembling a part, simulation results may be valid for this purpose. When the simulated object system is highly abstract, oversimplified, or in an area in which controversy exists over the body of knowledge—e.g., general management skills—decisions based on simulation results may make errors that seriously affect both employer and employee. Management games are ill-suited (in the author's opinion) for use in evaluating persons. However, some in-basket simulations can reveal whether supervisory candidates can notice interrelationships, establish priorities, delegate tasks, and seek expert advice rather than become bogged down in detail or ignore important issues. One company that uses simulations for evaluation purposes makes sure that each candidate experiences many simulations and has been seen for a long period of time by those who judge him. In other personnel applications, simulated jobs aid in evaluating rating systems, simulations of faked answers test questionnaires, and simulations of the decision processes of psychologists aid staffing the personnel department.

Physical Systems

All-computer simulation has been used to study object systems that are principally physical in nature. Examples are the bunching of charged-particle beams in an accelerator, the flow of underground water, the absorption of polymer molecules on solid surfaces, the behavior of ion fields, the loss effects in magnetic recording, the behavior of magnetic particles, the aging of geological systems, the behavior of water waves, weather forecasts, watersheds, and marine sedimentation.

Where control of physical systems by man is involved, both all-computer and man-computer simulations have been used to study object systems such as aircraft and spacecraft flight, man's role in managing water resources and controlling floods, long-range radar, and a variety of manufacturing plants. One researcher has been able to simulate tornadoes, but this has been done with special equipment that revolves air currents and is an example of machine simulation not covered in this book.

Continuous Systems

Since digital computers are discrete machines, there is a conceptual barrier to simulation of continuous systems. Most continuous system simulation is done with analog computers, which themselves are continuous systems. However, analog computers are not as accurate as digital computers so that all-computer simulation has found a place in representing continuous systems. An example is digital computer simulation of body movement during an auto crash. Some applications involve combinations of analog and digital computers to represent both the continuous object system and the management of it. An example is a simulated oil refinery.

Defense and Military

Simulation applications for studying military and defense systems are largely unpublished because of security requirements. These simulations range from economic and political models of international peace and war to representations of individual air and ground battles. Defense and military simulations involve all types of methods discussed in this book. In some cases the simulations are built into the operating systems, as in air defense networks, so that operators as a training exercise run the actual system but on simulated enemy attacks. Parts of the military-industrial complex that have been simulated are support planning, shipyards, logistics, long-range planning, procurement and distribution, information processing by teams, radar detection, tactical warfare, evaluations of equipment and organization, responses by damaged systems, material needs for complete theaters of military operations, weapons reliability, warhead payoff, over-the-beach landing and support operations, air combat, shipbuilding contracts, torpedo behavior, individual combat sorties, deployment strategies, equipment maintenance, mission reliability, and equipment production systems.

Flight and Space Simulation

The massive resource investment to place men on the moon has paid for substantial amounts of computer hardware and simulation programming. Simulation of the data in advance of a space launch provides a means to plan for handling data received from space during an actual flight fast enough to relay decisions back to the astronauts. Mathematical simulations of space flights—all-computer simulations without Monte Carlo features—have reduced the number of actual missile firings required to develop rockets.

Preparation for space flights requires many man-machine simulations, both for research and for training. Much of this work uses analog simulators not covered in this book. During space flights, the space capsule is simulated by another capsule on earth. All-computer simulations of both atmospheric and space flights can move systems of mathematical equations through time as fast as the object system they represent to provide environments for man-machine experimentation and training in "real time." An example is a computerized training cockpit that displays flight environments on a television screen.

Simulations of space capsule computers reveal the consequences of faults in the computer and permit study of ways to tell whether the onboard computer actually is faulting. Some models have simulated the behavior of the astronauts as they might use the onboard computer. At last report, these models can represent up to fourteen days of space flight.

EXERCISES

10.1. Read the definitions of a simulation model in at least three of the books listed under General Methodology in the bibliography. Compare these with the definition of a simulation model in Chapter 3. List the ideas common to all definitions. Can you explain the differences among the definitions, if any?

10.2. Read the descriptions of the ingredients and steps of simulation projects in at least three books listed under General Methodology in the bibliography. Do not select the same books read for Exercise 10.1. List the common ingredients and steps and explain differences, if any.

10.3. For a subject area of special interest to you, study a recent survey or review article, book of readings, or report of a symposium in which pertinent simulation applications are described. Write a brief paper stating types of simulations or simulation techniques described in this book that do *not* appear to be applicable to this subject area. Give reasons why they are not applicable.

10.4. For at least three of the periodicals listed in the bibliography, state the audience the periodical seems to be addressed to and state any differences you can infer among these audiences from the contents of the periodicals.

10.5. By examination of the literature for at least three categories in this chapter, find six reports of actual applications of simulation not mentioned anywhere in this book. Describe each briefly.

10.6. For any one of the above exercises you have already done, prepare a flow diagram for an all-computer model showing how you would simulate yourself carrying out the exercise.

10.7. Summarize the philosophical issues at stake when applying simulation methodology to practical problems where the consequences of error may be serious.

BIBLIOGRAPHY

The literature on simulation applications is vast; on methodology, varied and growing. For readers who wish to explore simulation applications and methodology further, this bibliography offers selected starting places. No attempt has been made to be exhaustive.

General Methodology

Chorafas, Dimitris N. *Systems and Simulation.* New York: Academic Press, 1965.

Evans, George W., II; Wallace, Graham F.; and Sutherland, Georgia L. *Simulation Using Digital Computers.* Englewood Cliffs, N. J.: Prentice-Hall, 1967.

Forrester, Jay W. *Industrial Dynamics.* Cambridge: M.I.T. Press; New York: John Wiley & Sons, 1961.

Gordon, Geoffrey. *System Simulation.* Englewood Cliffs, N. J.: Prentice-Hall, 1969.

Green, Bert F., Jr. *Digital Computers in Research, An Introduction for Behavioral and Social Scientists.* New York: McGraw-Hill Book Co., 1963.

Klerer, Melvin, and Korn, Granino. *Digital Computer User's Handbook.* New York: McGraw-Hill, 1967.

Ledley, Robert Steven. *Programming and Utilizing Digital Computers.* New York: McGraw-Hill, 1962.

Martin, Francis F. *Computer Modeling and Simulation.* New York: John Wiley & Sons, 1968.

Mayne, J. W. "Glossary of Terms Used in Gaming and Simulation." *Journal of Canadian Operational Research Society,* Vol. 4 (July, 1966): 114-118.

McLeod, John. *Simulation: The Dynamic Modeling of Ideas and Systems with Computers.* New York: McGraw-Hill, 1968.

McMillan, Claude, and Gonzalez, Richard F. *Systems Analysis,* Rev. ed. Homewood, Ill.: Richard D. Irwin, 1968.

Meier, Robert C.; Newell, William T.; and Pazer, Harold L. *Simulation in Business and Economics.* Englewood Cliffs, N. J.: Prentice-Hall, 1969.

Mize, Joe H., and Cox, J. Grady. *Essentials of Simulation.* Englewood Cliffs, N. J.: Prentice-Hall, 1968.

Naylor, Thomas H.; Balintfy, Joseph L.; Burdick, Donald S.; and Chu, Kong. *Computer Simulation Techniques.* New York: John Wiley & Sons, 1966.

Shapiro, George, and Rogers, Milton, eds. *Prospects for Simulation and Simulators of Dynamic Systems.* New York: Spartan Books, 1967.

"Simulation: Managing the Unmanageable." *SDC Magazine,* (April, 1965): 1.

Tocher, K. D. *The Art of Simulation.* London: The English Universities Press, 1963.

Symposia, Surveys, Reviews, and Readings

Abelson, Robert P. "Simulation of Social Behavior." In *Handbook of Social Psychology,* 2nd ed., Vol. 2, edited by Gardner Lindzey and Elliot Aronson. Reading, Mass.: Addison-Wesley Publishing Co., 1968.

Ackoff, Russell L., ed. *Progress in Operations Research.* Vol. 1, Chapters 9 and 10. Publications in Operations Research, No. 5. New York: John Wiley & Sons, 1963.

Alberts, W. E., and Malcolm, Donald G., eds. *Report of the Second*

System Simulation Symposium. Evanston, Ill.: American Institute of Industrial Engineers, Inc., 1960.

Burck, Gilbert. *The Computer Age*. New York: Harper & Row, 1965.

Chapman, R. L.; Kennedy, J. L.; Newell, A.; and Biel, W. C. "The Systems Research Laboratory's Air Defense Experiments." *Management Science*, Vol. 5 (1959): 250-269.

Cyert, Richard M. and March, James G. *A Behavioral Theory of the Firm*. Englewood Cliffs, N. J.: Prentice-Hall, 1963.

Fattu, Nicholas A., and Elam, Stanley, eds. *Simulation Models for Education*. Bloomington, Ind.: Phi Delta Kappa, 1965.

Feigenbaum, Edward A., and Feldman, Julian, eds. *Computers and Thought*. New York: McGraw-Hill, 1963.

Forrester, Jay W. "Industrial Dynamics—After the First Decade." *Management Science*, Vol. 14 (March, 1968): 398-414.

Geisler, Murray A. "Appraisal of Laboratory Simulation Experience." *Management Science*, Vol. 8 (1962): 239-245.

Geisler, M. A.; Haythorn, W. W.; and Steger, W. A. "Simulation and the Logistics Systems Laboratory." *Naval Research Logistics Quarterly*, Vol. 10 (March, 1963): 23-54.

Guetzkow, Harold, ed. *Simulation in Social Science: Readings*. Englewood Cliffs, N. J.: Prentice-Hall, 1962.

Hoggatt, Austin Curwood, and Balderston, Frederick E., eds. *Symposium on Simulation Models: Methodology and Applications to the Behavioral Sciences*. Cincinnati: South-Western Publishing Co., 1963.

Hollingdale, S. H., ed. *Digital Simulation in Operational Research*. New York: American Elsevier Publishing Co., 1967.

Loehlin, John C. *Computer Models of Personality*. New York: Random House, 1968.

Malcolm, Donald G., ed. *Report of the System Simulation Symposium*. New York: American Institute of Industrial Engineers, Inc., 1958.

Pattee, H. H.; Edelsack, E. A.; Fein, Louis; and Callahan, A. B., eds. *Natural Automata and Useful Simulations*. Washington, D.C.: Spartan Books, 1966.

Proceedings of the IBM Scientific Computing Symposium on Digital Simulation of Continuous Systems. White Plains, N. Y.: IBM Data Processing Division, 1967.

Proceedings of the IBM Scientific Computing Symposium on Simulation Models and Gaming. White Plains, N. Y.: IBM Data Processing Division, 1966.

Proceedings on Simulation in Business and Public Health. New York: American Statistical Association, 1969.

Rapoport, Anatol, and Orwant, Carol. "Experimental Games: A Review." *Behavioral Science*, Vol. 7 (1962): 1-37.

Record of Proceedings, First Annual Simulation Symposium. Tampa, Fla.: Annual Simulation Symposium, 1968.

Record of Proceedings, Second Annual Simulation Symposium. Tampa, Fla.: Annual Simulation Symposium, 1969.

"Simulation: A Symposium." *American Economic Review,* Vol. 50 (1960): 893-932.

Thomas, C. J., and Deemer, W. L., Jr. "The Role of Operational Gaming in Operations Research." *Operations Research,* Vol. 5 (1957): 1-27.

Tomkins, Silvan S., and Messick, Samuel, eds. *Computer Simulation of Personality.* New York: John Wiley & Sons, 1963.

Bibliographies

Abelson, Robert P. "Simulation of Social Behavior." In *Handbook of Social Psychology,* 2nd ed., Vol. 2, edited by Gardner Lindzey and Elliot Aronson. Reading, Mass.: Addison-Wesley Publishing Co., 1968.

Bibliography on Simulation. White Plains, N. Y.: IBM Corporation, 1966.

Deacon, Amos R. L., Jr. "A Selected Bibliography." In *Simulation and Gaming: A Symposium,* edited by Albert Newgarden. New York: American Management Association, 1961.

Duke, Richard D., and Schmidt, Allen H. "Operational Gaming and Simulation in Urban Research, An Annotated Bibliography." East Lansing, Mich.: Institute for Community Development and Services, Michigan State University, 1965.

Hartman, John J. "Annotated Bibliography on Simulation in the Social Sciences." Ames, Iowa: Iowa Agricultural and Home Economics Experiment Station, Iowa State University, 1966.

Malcolm, D. G. "Bibliography on the Use of Simulation in Management Analysis." *Operations Research,* Vol. 8 (1960): 169-177.

Naylor, Thomas H. "Bibliography 19. Simulation and Gaming." *Computing Reviews,* Vol. 10 (1969): 61-69.

"Selective Bibliography on Simulation Games as Learning Devices." *American Behavioral Scientist,* Vol. 10 (1966): 34-35.

Shubik, M. "Bibliography on Simulation, Gaming, Artificial Intelligence and Allied Topics." *Journal of the American Statistical Association,* Vol. 55 (1960): 736-751.

Werner, Roland, and Werner, Joan T. *Bibliography of Simulations: Social Systems and Education.* La Jolla, Calif.: Western Behavioral Sciences Institute, 1969.

Periodicals

American Behavioral Scientist: Sage Publications, Inc., Beverly Hills, Calif.
The American Economic Review: American Economic Association.
Behavioral Science: Mental Health Research Institute, University of Michigan, Ann Arbor, Mich.
Communications of the ACM: Association for Computing Machinery.
The Computer Journal: British Computer Society.
Computers and Automation: Berkeley Enterprises, Inc., Newtonville, Mass.
Computing Reviews: Association for Computing Machinery.
Computing Surveys: Association for Computing Machinery.
Conflict Resolution: Center for Research on Conflict Resolution, University of Michigan, Ann Arbor, Mich.
Datamation: F. D. Thompson Publications, Inc., Chicago, Ill.
The Gaming Newsletter: National Gaming Council, Washington, D.C.
IEEE Transactions on Systems Science and Cybernetics: Institute of Electrical and Electronics Engineers, Inc., New York, N. Y.
International Abstracts in Operations Research: International Federation of Operational Research Societies.
Journal of the ACM: Association for Computing Machinery.
Management Science: The Institute of Management Sciences, Providence, R. I.
Naval Research Logistics Quarterly: U. S. Office of Naval Research, Washington, D.C.
Operational Research Quarterly: Operational Research Society Ltd.
Operations Research: Operations Research Society of America.
Operations Research/Management Science (OR/MS): Executive Sciences Institute, Inc.
SICSIM (Special Interest Committee on Simulation) Newsletter: Association for Computing Machinery.
Simulation: Simulation Councils, Inc., San Diego, Calif.

Philosophical Issues

Conway, R. W.; Johnson, B. M.; and Maxwell, W. L. "Some Problems of Digital Systems Simulation." *Management Science,* Vol. 6 (1959): 92-110.

Conway, R. W. "Some Tactical Problems in Digital Simulation." *Management Science*, Vol. 10 (October, 1963): 47-61.

Crawford, Meredith P. "Dimensions of Simulation." *American Psychologist*, Vol. 21 (1966): 788-796.

Drabek, Thomas E., and Haas, J. Eugene. "Realism in Laboratory Simulation: Myth or Method?" *Social Forces*, Vol. 45 (1967): 337-346.

Frijda, Nico H. "Problems of Computer Simulation." *Behavioral Science*, Vol. 12 (1967): 59-67.

Hermann, Charles F. "Validation Problems in Games and Simulation with Special Reference to Models of International Politics." *Behavioral Science*, Vol. 12 (May, 1967): 216-231.

Naylor, Thomas H., and Finger, J. M. "Verification of Computer Simulation Models." *Management Science*, Vol. 14 (October, 1967): 92-101.

Simon, Herbert A. *The Sciences of the Artificial*. Cambridge, Mass.: M.I.T. Press, 1969.

Turing, A. M. "Computing Machinery and Intelligence." *Mind*, Vol. 59 (1950): 433-460. Reprinted in *The World of Mathematics*, Vol. 4, edited by James R. Newman. New York: Simon and Schuster, 1956, and in *Computers and Thought*, edited by Edward A. Feigenbaum and Julian Feldman. New York: McGraw-Hill, 1963.

Simulation Languages

Blunden, G. P., and Krasnow, Howard S. "The Process Concept as a Basis for Simulation Modeling." *Simulation*, Vol. 9 (August, 1967): 89-94.

Buxton, J. N., ed. *Simulation Programming Languages*. Amsterdam: North-Holland Publishing Co., 1968.

Forrester, Jay W. *Industrial Dynamics*. Cambridge: M.I.T. Press; New York: John Wiley & Sons, 1961.

Greenberger, Martin; Jones, Malcolm M.; Morris, James H., Jr.; and Ness, David N. *OnLine Computation and Simulation: The OPS-3 System*. Cambridge: M.I.T. Press, 1965.

Herscovitch, H., and Schneider, T. "GPSS III—An Expanded General Purpose Simulator." *IBM Systems Journal*, Vol. 4 (1965): 174-183.

Karr, Herbert W.; Kleine, Henry; and Markowitz, Harry M. *SIMSCRIPT 1.5*. Santa Monica, Calif.: California Analysis Center, Inc., 1965.

Kiviat, P. J., Villanueva, R., and Markowitz, H. M. *The SIMSCRIPT II Programming Language*. Englewood Cliffs, N. J.: Prentice-Hall, 1968.

Krasnow, Howard S., and Merikallio, Reino A. "The Past, Present, and Future of General Simulation Languages." *Management Science*, Vol. 11 (November, 1964): 236-267.

Krasnow, Howard S. "Computer Languages for System Simulation." In *Digital Computer User's Handbook*, edited by Melvin Klerer and Granino Korn. New York: McGraw-Hill, 1967.

Markowitz, Harry M.; Hausner, B.; and Karr, H. W. *SIMSCRIPT—A Simulation Language*. Englewood Cliffs, N. J.: Prentice-Hall, 1963.

McNeley, John. "Simulation Languages." *Simulation*, Vol. 9 (August, 1967): 95-98.

Newell, Allen; Tonge, Fred M.; Feigenbaum, Edward A.; Green, Bert F., Jr.; and Mealy, George H. *Information Processing Language*-V *Manual*, 2nd ed. Englewood Cliffs, N. J.: Prentice-Hall, 1964.

Pritsker, A. Alan B., and Kiviat, Philip J. *Simulation with GASP II*. Englewood Cliffs, N. J.: Prentice-Hall, 1969.

Pugh, Alexander L. *DYNAMO User's Manual*, 2nd ed. Cambridge, Mass.: M.I.T. Press, 1963.

SIMPAC User's Manual. TM-602/000/00, System Development Corporation, Santa Monica, Calif., 1962.

Teichroew, Daniel, and Lubin, John Francis. "Computer Simulation—Discussion of the Technique and Comparison of Languages." *Communications of the ACM*, Vol. 9 (1966): 723-741.

Teichroew, D.; Lubin, J. F.; and Truitt, Thomas D. "Discussion of Computer Simulation Techniques and Comparison of Languages." *Simulation*, Vol. 9 (October, 1967): 181-190.

Tocher, K. D. "Review of Simulation Languages." *Operational Research Quarterly*, Vol. 16 (1965): 184-204.

Monte Carlo Techniques

Gorenstein, S. "Testing a Random Number Generator." *Communications of the ACM*, Vol. 10 (1967): 111-118.

Hammersley, J. M., and Handscomb, D. C. *Monte Carlo Methods*. New York: John Wiley & Sons, 1965.

Hull, T. E., and Dobell, A. R. "Random Number Generators." *SIAM Review*, Vol. 4 (1962): 230-254.

Korn, Granino A. *Random-Process Simulation and Measurements*. New York: McGraw-Hill, 1966.

MacLaren, M. D., and Marsaglia, G. "Uniform Random Number Generators." *Journal of the ACM*, Vol. 12 (1965): 83-89.

Meyer, Herbert A., ed. *Symposium on Monte Carlo Methods.* New York: John Wiley & Sons, 1956.

Page, E. S. "The Generation of Pseudo-Random Numbers." In *Digital Simulation in Operational Research,* edited by S. H. Hollingdale. New York: American Elsevier Publishing Co., 1967.

Peach, P. "Bias in Pseudo-Random Numbers." *Journal of the American Statistical Association,* Vol. 56 (1961): 610-618.

Reference Manual: Random Number Generation and Testing. White Plains, N. Y.: IBM Data Processing Division, 1959.

Teichroew, Daniel. "A History of Distribution Sampling Prior to the Era of the Computer and Its Relevance to Simulation." *American Statistical Association Journal* (March, 1965): 27-49.

Shreider, Yu. A., ed. *Method of Statistical Testing: Monte Carlo Method.* Amsterdam: Elsevier Publishing Co., 1964.

Management Games

Cohen, Kalman J., and Rhenman, Eric. "The Role of Management Games in Education and Research." *Management Science,* Vol. 7 (January, 1961): 131-166.

Cohen, Kalman J.; Dill, William R.; Kuehn, Alfred A.; and Winters, Peter R. *The Carnegie Tech Management Game, An Experiment in Business Education.* Contributions to Management Education Series, Vol. 1. Homewood, Ill.: Richard D. Irwin, 1964.

Dale, Alfred G., and Klasson, Charles R. *Business Gaming, A Survey of American Collegiate Schools of Business.* Austin, Texas: Bureau of Business Research, The University of Texas, 1964.

Dill, William R.; Jackson, James R.; and Sweeney, James W., eds. *Proceedings of the Conference on Business Games.* New Orleans: Tulane University, 1961.

Greenlaw, Paul S.; Herron, Lowell W.; and Rawdon, Richard H. *Business Simulation in Industrial and University Education.* Englewood Cliffs, N. J.: Prentice-Hall, 1962.

Kibbee, Joel M.; Craft, Clifford J.; and Nanus, Burt. *Management Games, A New Technique for Executive Development.* New York: Reinhold Publishing Corp., 1962.

McKenney, James L. *Simulation Gaming for Management Development.* Boston: Division of Research, Graduate School of Business Administration, Harvard University, 1967.

Shubik, Martin. "Gaming: Costs and Facilities." *Management Science*, Vol. 14 (1968): 629-660.

Thorelli, Hans B., and Graves, Robert L. *International Operations Simulation, with Comments on Design and Use of Management Games*. New York: Free Press of Glencoe, 1964.

Special Techniques

Barton, Richard F. "A Generalized Responsiveness (Elasticity) Function for Simulations." *Behavioral Science*, Vol. 12 (1967): 337-343.

Gafarian, A. V., and Ancker, C. J. "Mean Value Estimation from Digital Computer Simulation." *Operations Research*, Vol. 14 (1966): 25-44.

Giffen, Sidney F. *The Crisis Game, Simulating International Conflict*. Garden City, N. Y.: Doubleday & Company, 1965.

Guetzkow, Harold; Alger, Chadwick F.; Brody, Richard A.; Noel, Robert C.; and Snyder, Richard C. *Simulation in International Relations: Developments for Research and Teaching*. Englewood Cliffs, N. J.: Prentice-Hall, 1963.

Naylor, T. H.; Wertz, Kenneth; and Wonnacott, Thomas. "Methods for Analyzing Data from Computer Simulation Experiments." *Communications of the ACM*, Vol. 10 (1967): 703-710.

Orcutt, Guy H.; Greenberger, Martin; Korbel, John; and Rivlin, Alice M. *Microanalysis of Socioeconomic Systems: A Simulation Study*. New York: Harper & Row, 1961.

Orr, William D., ed. *Conversational Computers*. New York: John Wiley & Sons, 1968.

Pool, I.; Abelson, R. P.; and Popkin, S. *Candidates, Issues, and Strategies: A Computer Simulation of the 1960 Presidential Election*. Cambridge, Mass.: M.I.T. Press, 1964.

Sass, Margo, and Wilkinson, W. D. *Computer Augmentation of Human Reasoning*. Washington, D.C.: Spartan Books, 1965.

INDEX